Applied Dynamics

Werner Schiehlen • Peter Eberhard

Applied Dynamics

Springer

Werner Schiehlen
Institute of Engineering
and Computational Mechanics
University of Stuttgart
Stuttgart, Germany

Peter Eberhard
Institute of Engineering
and Computational Mechanics
University of Stuttgart
Stuttgart, Germany

This is the revised and updated English translation of the third edition of the German book Technische Dynamik originally published in 1986 by Vieweg + Teubner.

ISBN 978-3-319-37457-4 ISBN 978-3-319-07335-4 (eBook)
DOI 10.1007/978-3-319-07335-4
Springer Cham Heidelberg New York Dordrecht London

Printed on acid-free paper

Springer is part of Springer Science+Business Media (www.springer.com)

Preface

This book *Applied Dynamics* is the revised English edition of the third edition of the German book *Technische Dynamik* originally published in 1986. The second and third editions have been co-authored by Peter Eberhard, too. The present English edition was encouraged by Nathalie Jacobs from Springer-Verlag, who deserves our sincere thanks. A draft translation of the third edition was prepared by Aaron Kuchle. Then, the authors revised the book and added to the many German references appropriate English books and papers, too.

In the last decades, applied dynamics has been incorporated into many new areas beyond its original applications in machine and vehicle dynamics. Models of robot dynamics are being successfully exploited in biomechanics, three-dimensional hinges are helping in the development of middle ear protheses, and contact models for a few bodies are also being utilized for systems of a great number of particles. At the same time, model-based design of control-systems requires low-dimensional systems as provided by multibody dynamics. Complex models with finite elements can be represented with new methods of model reduction under given error limits. Complex models, also containing tire models of vehicles, are also being used, e.g., in the control of self-balancing Segway scooters. Moreover, efficient models are indispensable for the rapidly developing field of simulation technology.

All these new challenges and methods require extensive knowledge of the foundations of applied dynamics which remain just as relevant as ever for academic education. Lectures at universities as well as continuing education in industry have the essential task in common, namely the axiomatic, computer-based modelling of mechanical and mechatronic systems, what requires advanced applied dynamics. The structure and organization of the book have proved to be very effective, so these have been kept unchanged in the English edition. Moreover, we preserved the *43 Examples* completely integrated in the 8 Chapters of the book. These *Examples* show educational applications of the fundamentals presented but they can be also used as problems for self-study. However, the opportunity was taken to correct printing errors, to revise notation, and to undertake many other alterations. In particular, the section on Beam System has been renamed to Flexible Systems, and it allows the reader access to further literature including the Floating Frame of References Formulation and the Absolute Nodal Coordinate Formulation.

We would like to thank our many attentive readers, our colleagues and coworkers at the Institute for Engineering and Computational Mechanics as well as our students for their suggestions and inquiries, which have exerted great influence in the revision of the manuscript and the changes of the drawings. We are pleased to receive comments and notifications of possible errors, which can never be completely avoided even after careful proofreading and which will be continuously documented on the book's website.[1] We hope that this book will continue to be useful in education and in engineering practice, and we wish all interested readers much success and satisfaction in their engagement with this fascinating topic.

Stuttgart, Germany
January 2014

Werner Schiehlen
Peter Eberhard

[1] www.itm.uni-stuttgart.de/book_applied_mechanics

From the Preface to the First German Edition

This book emanated from the thankworthy encouragement of my revered teacher, Prof. Dr. Kurt Magnus. It is based on lectures on applied dynamics and machine dynamics given at the Technical University of Munich and the University of Stuttgart as well as research on robot dynamics during a sabbatical with the MAN New Technology Division in Munich.

Applied dynamics, a branch of engineering mechanics, is presently a widely ramified science with applications in machine and vehicle engineering, aerospace, and even control technology. An introductory textbook can therefore only present the foundations as well as selected examples. Yet it is a concern of this book, mainly written for engineers, to present the computational methods in use today on a common basis. For this purpose, we use analytical mechanics as a base, whereby the Lagrange version of the d'Alembert principle has proved to be particularly productive. It is thus possible to discuss the method of multibody systems, the method of finite elements, and the method of continuous systems in a consistent manner. This makes it possible for the student to reach a deeper understanding with less effort. Also, the practicing engineer will be in a better position to assess his calculation results.

The book is divided into nine chapters. In the introduction, the problem of modelling is discussed, while the second chapter is dedicated to kinematics. The basics of kinematics are described at great detail, since they are necessary not only in kinetics but also for the principles of analytical mechanics. In the third chapter, the basic ideas of kinetics are presented for the mass point, the rigid body, and the continuum. This is followed by the principles of mechanics in Chap. 4, where only those important for engineering applications will be discussed. Chapters 5–7 then deal in turn with multibody systems, finite element systems, and continuous systems. The equations of motion are converted in Chap. 8 into state equations, which are consistent for all mechanical systems. The ninth chapter deals with a few questions regarding numerical solution methods.

...

Stuttgart, Germany
Fall 1984

Werner Schiehlen

Contents

Chapter 1
Introduction

Applied dynamics is concerned with the characteristics of motion and deformation and is based on kinematics, kinetics, and the principles of analytical mechanics. Mechanical systems are usually engineering structures. In order to investigate these mathematically, it is necessary to describe them using models. According to the kind of modelling, a distinction is made in this book between multibody systems, finite element systems, and continuous systems. All of these mechanical models lead from their equations of motion to state equations, which can be solved numerically by consistent methods.

Applied dynamics arose from the classical machine dynamics of power machines. Today it also comprises biomechanics, structural dynamics, vehicle dynamics, robot dynamics, rotor dynamics, satellite dynamics, and many areas of system dynamics. One common basis of all these independent disciplines are the mechanical systems, whose modelling is always the first stage of their engineering and scientific investigation.

1.1 Tasks of Applied Dynamics

Concerning the tasks of applied dynamics, it remains still valid today what Biezeno and Grammel [10] wrote in the year 1939 in the preface of their homonymous book:

> In the organization and treatment of the subject matter, we constantly bore in mind that for technology a problem is only worth solving if it has the possibility of practical application, and that an engineering problem can only be considered as solved if the solution can be evaluated numerically in every detail with tolerable calculation effort.

With this in mind, applied dynamics represents an important branch of mechanics that can no longer do without the use of computers and thus also belongs to the field of 'computational mechanics'.

W. Schiehlen and P. Eberhard, *Applied Dynamics*, DOI 10.1007/978-3-319-07335-4_1,

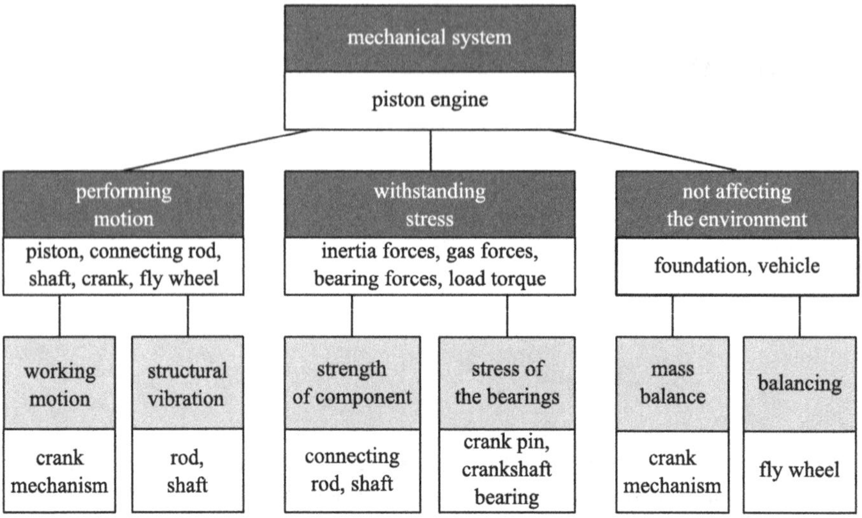

Fig. 1.1 Tasks of applied dynamics using the example of a piston engine

The tasks of applied dynamics result directly from the engineer's practical needs. A mechanical system should often carry out motions while withstanding stresses and not burdening the environment. Figure 1.1 shows some possible problems using the example of a piston engine.

In order to solve these problems, first the equations of motion and equations of reaction of mechanical systems are required. These can be obtained with the help of analytical mechanics.

1.2 Contributions of Analytical Mechanics

The equations of motion of free mechanical systems have been known since the early days of mechanics. Newton (1643–1727) published in 1687 his three famous basic laws: the law of inertia, the law of motion, and the law of reaction. The law of motion provides directly the equations of motion of a mass point. Euler (1707–1783) made the equations of motion available for a rigid body in 1775 with the principle of linear and angular momentum. D'Alembert (1717–1783) published his principle for constrained point systems in 1743, which Lagrange (1736–1813) formulated more simply in the year 1788 using the principle of virtual work. In particular, Lagrange introduced generalized coordinates, which are also the basis of his equations of motion of the second kind of 1811. The principle of Gauss (1777–1855), published in 1829, and the principle of Jourdain of 1908 represent generalizations of the d'Alembert principle. The Lagrange equations of the second kind were extended to nonholonomic constrained systems in 1879 by Gibbs and in 1900 by Appell. In addition to the previously mentioned differential principles, the principle of Hamilton (1805–1865) published in 1834 should also be mentioned as an integral principle. Further information about the historical development of these principles can be found in Szabo [63]. All the important principles of mechanics have been compiled by Päsler [41]. Detailed discussions of analytical mechanics are found in Budo [13] and Hamel [25]. Arnold [1] and Papastavridis [40] provide a more modern approach to classical mechanics.

1.3 Modelling Mechanical Systems

Mechanical systems are always characterized by components with inertia and elasticity. The influences of damping and excitation by external forces are usually also involved, see Fig. 1.2. The inertia of a component is determined by its volume and density. The mass can be affected by the dimensions and density of a component. It is always positive and is assumed to be temporally unchanging. The elasticity of a component depends on its geometric shape and material properties. By means of a suitable design, it is possible to obtain an elasticity that is more important than the mass. We then speak of spring elements, such as e.g. leaf springs, helical springs, and torsion springs. Irretier [28] or Demeter [15] provide a nice overview of this. Damping can be generated either by material damping in the components, by friction phenomena between moving components, or by constructively shaped damping elements. External forces can be caused on the one hand by the effect of force fields like gravitation and by special drive elements such as actuators. On the other hand, they can also be a reaction to a motion given by position actuators, due to joints for example.

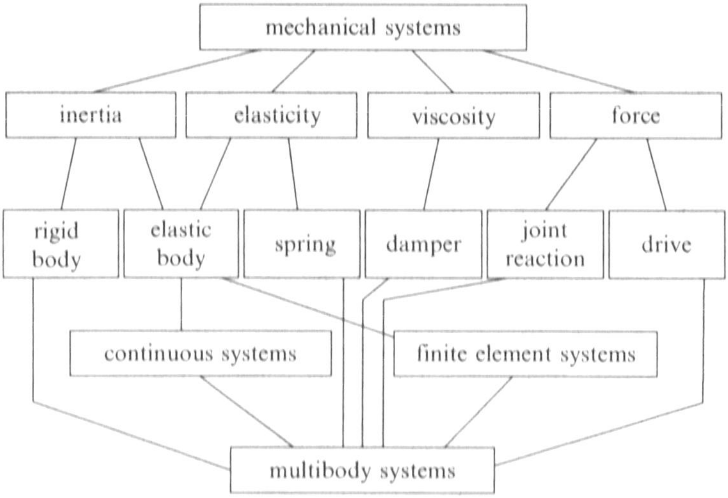

Fig. 1.2 Properties of mechanical systems

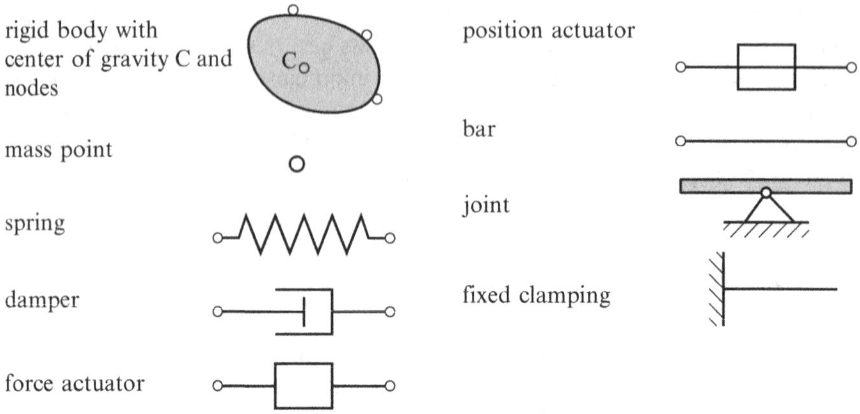

Fig. 1.3 Elements of a multibody system

The properties of a real engineering system must now be described with idealized models. In this process, we differentiate between models with distributed and those with concentrated parameters. To the models with distributed parameters belongs in particular the elastic body of continuum mechanics. Models with concentrated parameters are found in stereo-mechanics. They comprise rigid bodies, massless springs, massless dampers, as well as driving and reaction forces. Mechanical models can then be constructed from these assumptions.

1.3.1 Multibody Systems

A multibody system is composed of rigid bodies with mass, upon which act concentrated forces and torques at discrete points. The forces and torques originate from massless springs, dampers, and actuators as well as inflexible joints and other kinds of bearings. In addition, applied volume forces and torques can act on the rigid bodies. Frequently used symbols for the elements of a multibody system are compiled in Fig. 1.3.

The symbol of the rigid body characterizes its mass inertia. Characteristic points of the rigid body are the center of mass C and a finite number of nodes, at which concentrated forces and torques are acting. In the special case of a mass point, all the nodes coincide and the moments of inertia vanish. The spring symbol is reminiscent of a helical spring, yet a special case of the spring is also the massless bar, which appears e.g. in elastic frameworks. The damper symbol is inspired by a hydraulic damper, but it should also be used for electric and magnetic dampers. In the case of actuators, we distinguish between force actuators, which develop forces, and position actuators, which initiate motion. A blocked position actuator corresponds to a rigid bar, which is obtained in the limit case from an infinitely stiff spring, too. Bearings are assumed to be rigid, i.e. lacking deformations in locked bearing directions, while ideal bearings are also free of friction. The massless elements can also be subdivided according to the type of forces into coupling elements (springs, dampers, force actuators) and bearing elements (bars, joints, position actuators). The former generate applied forces, the latter reaction forces.

The method of multibody systems is based on the fact that the properties of inertia, elasticity, damping, and force are assigned to particular discrete elements. The individual, locally described elements are then assembled under consideration of bearings in a global system. Due to the discretization, we obtain the comparatively simple global equations of motion, which describe the mechanical system for the selected idealizations and approximations. Many details concerning problems of multibody dynamics can also be found in Rill and Schäffer [45], Bauchau [7], Woernle [66], or also in Popp and Schiehlen [43].

Mass point systems are a special case of the multibody system and have been known for a very long time in mechanics. Mass point systems have been investigated systematically by classical analytical mechanics, resulting in a large body of knowledge on the topic, see e.g. Hiller [26]. Interest in multibody systems only began to heighten after 1965 as a response of the corresponding requirements of space travel. Since then, computer-based formalisms have also been developed. Multibody systems, as opposed to mass point systems, also allow a simple treatment of gyroscopic phenomena.

1.3.2 Finite Element Systems

A finite element system consists of an arrangement of deformable elements or subsystems with inertia, upon which concentrated forces and torques at discrete points called "nodes" are acting. Applied surface or volume forces can also be involved. The bearing of finite element systems takes place at the nodes. Some frequently used elements can be seen in Fig. 1.4. Bars and beams are one-dimensional elements, while the triangle is a two-dimensional element and the cuboid an example of a three-dimensional volume element.

The basic idea behind the finite element method is that the properties of inertia, elasticity, and force are taken into account in a discrete element with a simple geometry. For this reason, one must first determine the local equations of motion of a single finite element. From the individual finite elements, the global system is then assembled by linking the nodes. From the local equations of motion we are then generating the global equations of motion. Because of the introduction of discrete elements and the selected shape functions within an element, the finite element method represents an approximation method, as does the method of multibody systems.

Elastic frameworks, a special case of the finite element system, have been thoroughly considered in the past in the context of elastostatics. The high order of the mostly linear systems of equations involved has made its engineering application difficult for many decades. The method of finite elements had its breakthrough at

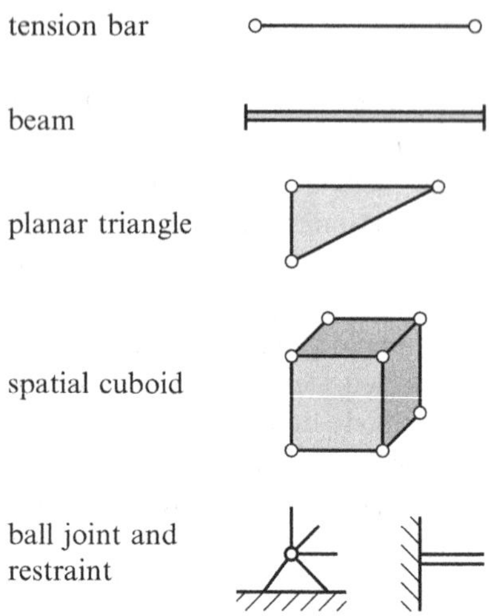

Fig. 1.4 Some finite elements and bearings

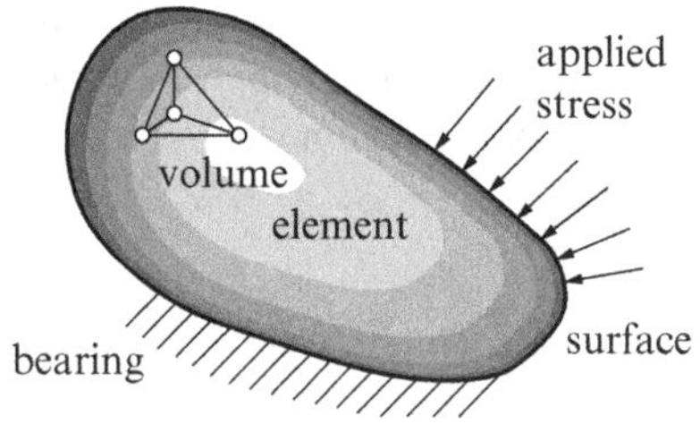

Fig. 1.5 Elastic body with an infinitesimal element

the end of the 1950s when the development of computer systems also permitted the solution of elaborate systems of equations. Nowadays, numerous verified program systems are available which make it possible to discretize and to solve continuum problems of structural dynamics automatically.

1.3.3 Continuous Systems

A continuous system consists of elastic bodies with mass in whose volumes continuously distributed applied forces are acting, and on whose surface impinge continuously distributed forces (stresses) are found. The surface stresses originate either from applied stresses or constraint stresses as a result of a given bearing, see Fig. 1.5.

Modelling with continuous systems is based on the continuous distribution of mass and elasticity in the body. The equations of motions can therefore only be formulated for an infinitesimally small volume element, representing partial differential equations dependent on space and time. In contrast to the methods of

Table 1.1 Models of mechanical systems

Mechanical model	Geometric shape	Stiffness distribution
Multibody system	Complicated	Inhomogeneous
Finite element system	Complicated	Homogeneous
Continuous system	Simple	Homogeneous

multibody systems and finite elements, continuous systems are dealt with exactly in terms of continuum mechanics. A strict solution of the local equations of motion is however only possible in simple cases, e.g. with bars and beams. In the general case, the numerical solution requires a mathematical discretization, so ultimately continuous systems also involve approximation methods. Yet for engineering praxis, mechanical discretization is often more intuitive and easier, which is the basis for the considerable success of discrete mechanical systems.

1.3.4 Flexible Multibody Systems

We refer to multibody systems as flexible when rigid and elastic bodies are both employed for modelling a mechanical system. Difficulties in modelling can arise at the interface between a rigid and an elastic body, but the engineer can master these by additional assumptions. Applications of flexible multibody systems are known in vehicle, robot, and satellite dynamics amongst other areas. The beam systems discussed in Sect. 6.3 also belong to the category of flexible multibody systems. A detailed description can be found in Schwertassek and Wallrapp [51], Shabana [56, 57], or Gerardin [22].

1.3.5 Selecting a Mechanical Model

The choice of a suitable mechanical model requires a lot of experience. General points of reference for modelling are the elastic stiffness distribution and the geometric shape of the given engineering system, see Table 1.1. For the elastic degree of freedom, the method of multibody systems generally provides low, the finite element method high eigenfrequencies, see Sect. 9.3.

Example 1.1 (Single-Cylinder Engine). For the single-cylinder engine shown in Fig. 1.6, the following problems are possible subjects of investigation:

1. Motion of the crank mechanism
2. Bending vibrations of the connecting rod
3. Torsion vibrations of the shaft.

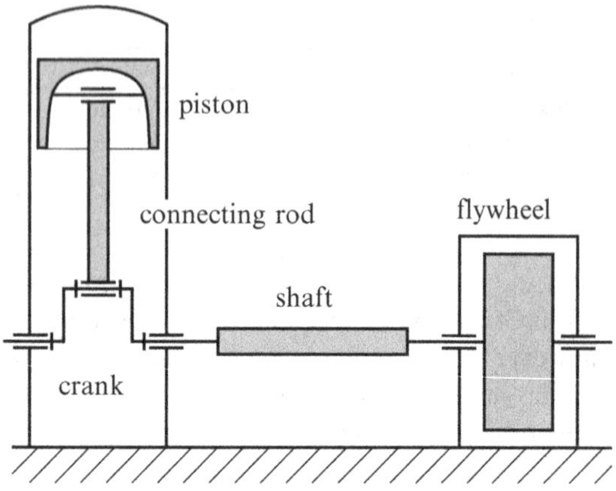

Fig. 1.6 Single-cylinder engine with flywheel

In order to solve the first problem, the piston, connecting rod, crank, and flywheel are modelled together as a multibody system. The second problem is solved with a finite element system, since the connecting rod exhibits a complicated geometric shape. The third problem can be investigated by means of a continuous system.

End of Example 1.1.

1.3.6 Number of Degrees of Freedom

The number of degrees of freedom plays a fundamental role in all mechanical systems regardless of the type of modelling. Both modelling accuracy and computation cost increase with the number of degrees of freedom. This means that determining the number of degrees of freedom is a true task of an engineer, which can usually only be achieved in conjunction with a technically sensible compromise. The lower limit for the degrees of freedom is given by the degrees of freedom of the rigid body, while the upper limit is infinity in the case of an elastic body. As a rule, the degrees of freedom of the rigid body are supplemented by a finite number of elastic degrees of freedom, thus choosing a compromise. Some disciplines can even do without such a compromise. In classical machine dynamics, mainly the rigid body's degrees of freedom are taken into consideration, while in structural dynamics the rigid body's degrees of freedom are dropped and a large number of elastic degrees of freedom are utilized.

The number e of degrees of freedom of a free mechanical system with p elements is obtained from the number e_i of degrees of freedom of the individual elements according to the relation

$$e = \sum_{i=1}^{p} e_i. \tag{1.1}$$

Table 1.2 Degrees of freedom e_i of a free element

Type of element	1-dimensional	2-dimensional	3-dimensional
Mass point	1	2	3
Rigid body	1	3	6
Finite beam element	$2 \cdot 1$	$2 \cdot 3$	$2 \cdot 6$
Finite tetrahedral element	$2 \cdot 1$	$3 \cdot 2$	$4 \cdot 3$
Finite cube element	$2 \cdot 1$	$4 \cdot 2$	$8 \cdot 3$
Mass point of a nonpolar continuum	1	2	3
Mass point of a polar continuum	1	3	6

For one-dimensional, two-dimensional, and three-dimensional problems, the numbers given for e_i in Table 1.2 are applicable. Thus e.g., $e = 6p$ is just as valid for a single free three-dimensional rigid body as for a single planar tetrahedral element. The number of elements depends in turn upon the discretization.

In the assembly of the global system, p elements are linked via q independent constraints or bearings. The number f of positional degrees of freedom of a constrained mechanical system is then only

$$f = e - q. \tag{1.2}$$

Based on the foundations of analytical mechanics, applied dynamics makes it possible to obtain f differential equations for the motion and q algebraic equations for the reaction forces. Copious use will be made of this in the following chapters.

In the mass point of a nonpolar continuum belonging to a volume element, only displacements can occur, while in polar continua twisting can also be found. Analogously, only forces or stresses are active in nonpolar continua, while torque stresses must also be considered in the case of polar continua, also known as Cosserat continua.

Table 1.2 shows a relationship which is of great methodical interest. The number of degrees of freedom of the mass points of a point system and the mass points of a nonpolar continuum on the one hand, and the number of degrees of freedom of rigid bodies and a polar continuum, on the other hand, coincide with each other. This relationship, which was also discussed by Schäfer [48] in his detailed report about the Cosserat continuum, will come up again in Sect. 4.1. The monograph of Rubin [46] also deals with these.

Chapter 2
Basic Kinematics

In applied dynamics, we draw a distinction between free systems with elements that can move without restrictions and constrained systems, the elements of which are bound to each other or their surroundings by means of bearings. While, for example, satellite dynamics is largely concerned with free systems, machine dynamics deals almost solely with constrained systems. This chapter will provide a summary of the kinematic principles behind mass point systems, multibody systems, and continuous systems. From a kinematic standpoint, finite element systems are considered as continuous systems and will therefore not be treated separately. The kinematics of free and constrained systems will be presented both in a spatially fixed inertial frame as well as in a frame in relative motion.

2.1 Free Systems

Free mechanical systems have especially simple kinematics since their motion is not subject to constraints from any kind of bearings. Their mathematical description is first undertaken with reference to a spatially fixed frame, though often generalized coordinates are used in addition to Cartesian coordinates.

2.1.1 Kinematics of a Mass Point

The mass point is the simplest model in mechanics. However, a single free point has little engineering importance. Free mass point systems on the other hand can be encountered in flying elastic structures (e.g. in aerospace engineering) or in systems containing only coupling elements instead of bearings. Moreover, any elastic continuum can be seen as a free system of infinitely many mass points. In free systems, all points are kinematically equal. For this reason, we will first examine the single free point in some detail.

W. Schiehlen and P. Eberhard, *Applied Dynamics*, DOI 10.1007/978-3-319-07335-4_2,

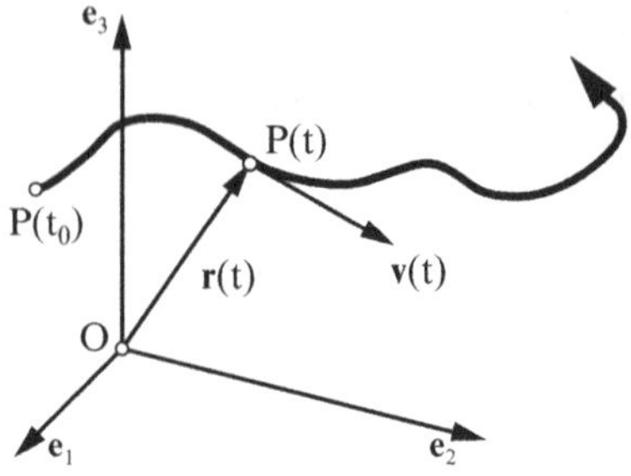

Fig. 2.1 Free motion of a mass point

The current position of a point in motion $P(t)$ at time t in space is clearly described with respect to the origin O of the fixed frame by means of the location vector $\boldsymbol{r}(t)$, see Fig. 2.1. In the course of time, the point in motion P changes its location, following the path marked by the position vector $\boldsymbol{r}(t)$. This motion is called displacement or translation.

Every location vector can be clearly resolved into its components in a Cartesian frame $\{O;\boldsymbol{e}_\alpha\}$, $\alpha = 1(1)3$, with the origin O and the basis vectors $\boldsymbol{e}_\alpha$. The current position is then defined for the location vector as

$$\boldsymbol{r}(t) = r_1(t)\boldsymbol{e}_1 + r_2(t)\boldsymbol{e}_2 + r_3(t)\boldsymbol{e}_3. \tag{2.1}$$

In a given frame, the location vector $\boldsymbol{r}(t)$ can thus be clearly represented, using (2.1), by the 3×1 vector of its coordinates

$$\boldsymbol{r}(t) = \begin{bmatrix} r_1 \; r_2 \; r_3 \end{bmatrix}. \tag{2.2}$$

The coordinates are generally written without an argument, and no distinction is made between row and column vectors, see Sect. A.2 and (A.34).

A free point in space has three degrees of freedom. Three coordinates are required to describe these. In addition to the Cartesian coordinates $r_\alpha, \alpha = 1(1)3$, from (2.2), we can also make use of generalized coordinates $x_\gamma, \gamma = 1(1)3$, which are as a rule curvilinear. These generalized coordinates can then be merged into a 3×1 position vector

$$\boldsymbol{x}(t) = \begin{bmatrix} x_1 \; x_2 \; x_3 \end{bmatrix}. \tag{2.3}$$

In general, there is a nonlinear correlation between the location vector $\boldsymbol{r}(t)$ and the position vector $\boldsymbol{x}(t)$,

$$\boldsymbol{r}(t) = \boldsymbol{r}(\boldsymbol{x}(t)) = \boldsymbol{r}(\boldsymbol{x}), \tag{2.4}$$

which in some cases simplifies the description of a point motion considerably. For example, circular motions can be represented more clearly by cylindrical coordinates than with Cartesian coordinates. Another example worth mentioning are the spatial central forces, which in spherical coordinates have only one nonzero coordinate. Some information on the notation used in this book can be found in Sect. A.1.

The velocity $\boldsymbol{v}(t)$ of point P is obtained by differentiation of (2.2) according to time, while its direction is determined by the tangent to the path, see Fig. 2.1. In a fixed frame (inertial frame), the 3×1 vector of the absolute velocity is thus written

$$\boldsymbol{v}(t) = \dot{\boldsymbol{r}}(t) = \begin{bmatrix} \dot{r}_1 \; \dot{r}_2 \; \dot{r}_3 \end{bmatrix}, \tag{2.5}$$

where $\dot{\boldsymbol{r}}$ is the derivative of $\boldsymbol{r}$ with respect to time t. The velocity can also be expressed in generalized coordinates. From (2.4) and (2.5), we obtain according to the chain rule

$$\boldsymbol{v}(t) = \boldsymbol{v}(\boldsymbol{x},\dot{\boldsymbol{x}}) = \frac{\partial \boldsymbol{r}}{\partial \boldsymbol{x}} \cdot \frac{d\boldsymbol{x}}{dt} = \boldsymbol{H}_T(\boldsymbol{x}) \cdot \dot{\boldsymbol{x}}(t), \tag{2.6}$$

whereby we obtain the 3×3 Jacobian matrix of translation

$$\boldsymbol{H}_T(\boldsymbol{x}) = \begin{bmatrix} \frac{\partial r_1}{\partial x_1} & \frac{\partial r_1}{\partial x_2} & \frac{\partial r_1}{\partial x_3} \\ \frac{\partial r_2}{\partial x_1} & \frac{\partial r_2}{\partial x_2} & \frac{\partial r_2}{\partial x_3} \\ \frac{\partial r_3}{\partial x_1} & \frac{\partial r_3}{\partial x_2} & \frac{\partial r_3}{\partial x_3} \end{bmatrix}, \tag{2.7}$$

which establishes a relation between the location vector and the generalized coordinates. The velocity is therefore a linear function of the first time derivative $\dot{\boldsymbol{x}}(t)$ of the selected position vector.

The functional or Jacobian matrices are very important in applied dynamics. Their underlying mathematical principles are found in the differential and integral calculus of functions of several variables, see e.g. Bronstein and Semendjajew [12]. Since defining the Jacobian matrices element by element via scalar differential quotients is time-consuming, we shall resort to matrix notation. The 3×3 Jacobian matrix (2.7) in the form

$$\boldsymbol{H}_T(\boldsymbol{x}) = \frac{\partial \boldsymbol{r}(\boldsymbol{x})}{\partial \boldsymbol{x}} \tag{2.8}$$

thus follows from the 3×1 vector $\boldsymbol{r}(\boldsymbol{x})$ of the dependent variables and the 3×1 vector $\boldsymbol{x}$ of the independent variables, see (A.36). In this notation, the following relation is generally true for an $e \times 1$ vector $\boldsymbol{x}$

$$\frac{\partial \boldsymbol{x}}{\partial \boldsymbol{x}} = \boldsymbol{E}, \tag{2.9}$$

where $\boldsymbol{E}$ is the $e \times e$ unit matrix. Also, for the $e \times 1$ vectors $\boldsymbol{r}$ and $\boldsymbol{x}$, we obtain a result corresponding to (2.9)

$$\frac{\partial \boldsymbol{r}(\boldsymbol{x}(\boldsymbol{r}))}{\partial \boldsymbol{r}} = \frac{\partial \boldsymbol{r}}{\partial \boldsymbol{x}} \cdot \frac{\partial \boldsymbol{x}}{\partial \boldsymbol{r}} = \boldsymbol{E}. \tag{2.10}$$

In addition, the chain rule is written with an additional $f \times 1$ vector $\mathbf{y}$ as follows,

$$\frac{\partial \boldsymbol{r}(\boldsymbol{x}(\boldsymbol{y}))}{\partial \boldsymbol{y}} = \frac{\partial \boldsymbol{r}}{\partial \boldsymbol{x}} \cdot \frac{\partial \boldsymbol{x}}{\partial \boldsymbol{y}}, \tag{2.11}$$

yielding an $e \times f$ matrix. The computer-friendly notation introduced here will be used continually in the following. It also contains, for $e = 3$, the relations of the vector analysis.

The acceleration $\boldsymbol{a}(t)$ of point P is a measure of the change in time of its velocity and is determined by differentiation of (2.5) with respect to time. In a spatially fixed frame, the 3×1 vector of the absolute acceleration coordinates is thus defined by

$$\boldsymbol{a}(t) = \dot{\boldsymbol{v}}(t) = \ddot{\boldsymbol{r}}(t) = \begin{bmatrix} \ddot{r}_1 \; \ddot{r}_2 \; \ddot{r}_3 \end{bmatrix}. \tag{2.12}$$

Acceleration can be expressed not only with Cartesian coordinates using (2.12), but also with generalized coordinates. With the product rule, (2.6) yields the relation

$$\begin{aligned} \boldsymbol{a}(t) = \boldsymbol{a}(\boldsymbol{x},\dot{\boldsymbol{x}},\ddot{\boldsymbol{x}}) &= \boldsymbol{H}_T(\boldsymbol{x}) \cdot \ddot{\boldsymbol{x}}(t) + \frac{d\boldsymbol{H}_T(\boldsymbol{x})}{dt} \cdot \dot{\boldsymbol{x}}(t) \\ &= \boldsymbol{H}_T(\boldsymbol{x}) \cdot \ddot{\boldsymbol{x}}(t) + \left(\frac{\partial \boldsymbol{H}_T(\boldsymbol{x})}{\partial \boldsymbol{x}} \cdot \dot{\boldsymbol{x}}(t) \right) \cdot \dot{\boldsymbol{x}}(t). \end{aligned} \tag{2.13}$$

The acceleration is thus a linear function of the second derivative $\ddot{\boldsymbol{x}}(t)$ of the position vector. Moreover, it is also quadratically dependent on the first derivative $\dot{\boldsymbol{x}}(t)$ of the position vector.

All equations essential for the kinematics of a point have thereby been established.

Example 2.1 (Point Movement in Spherical Coordinates). For problems with centrally symmetric forces, the spherical coordinates ψ, ϑ, R, as shown in Fig. 2.2, are often to be recommended. The 3×1 position vector is then

$$\boldsymbol{x}(t) = \begin{bmatrix} \psi \; \vartheta \; R \end{bmatrix}. \tag{2.14}$$

The 3×1 location vector then acquires the form

$$\boldsymbol{r}(t) = \begin{bmatrix} \cos\psi \sin\vartheta \\ \sin\psi \sin\vartheta \\ \cos\vartheta \end{bmatrix} R \tag{2.15}$$

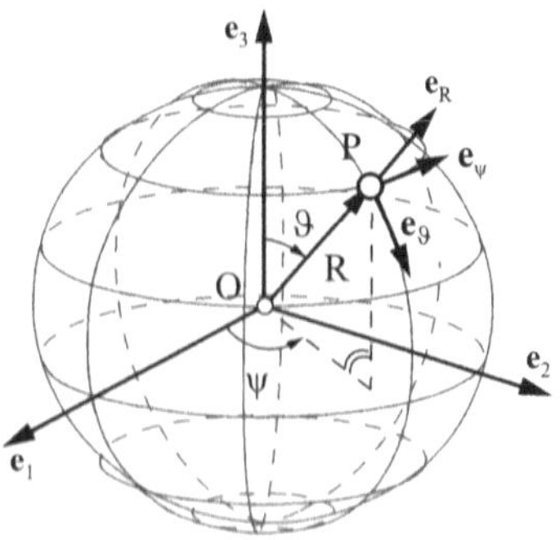

Fig. 2.2 Spherical coordinates

and the 3×3 Jacobian matrix of translation is written in accordance with (2.7) or (2.8)

$$\boldsymbol{H}_T(\boldsymbol{x}) = \begin{bmatrix} -R\sin\psi\sin\vartheta & R\cos\psi\cos\vartheta & \cos\psi\sin\vartheta \\ R\cos\psi\sin\vartheta & R\sin\psi\cos\vartheta & \sin\psi\sin\vartheta \\ 0 & -R\sin\vartheta & \cos\vartheta \end{bmatrix}. \tag{2.16}$$

With this, the 3×1 velocity vector is also determined with (2.6),

$$\boldsymbol{v}(\boldsymbol{x},\dot{\boldsymbol{x}}) = \begin{bmatrix} -R\dot{\psi}\sin\psi\sin\vartheta + R\dot{\vartheta}\cos\psi\cos\vartheta + \dot{R}\cos\psi\sin\vartheta \\ R\dot{\psi}\cos\psi\sin\vartheta + R\dot{\vartheta}\sin\psi\cos\vartheta + \dot{R}\sin\psi\sin\vartheta \\ -R\dot{\vartheta}\sin\vartheta + \dot{R}\cos\vartheta \end{bmatrix} \tag{2.17}$$

yielding the acceleration vector

$$\boldsymbol{a}(\boldsymbol{x},\dot{\boldsymbol{x}},\ddot{\boldsymbol{x}}) =$$

$$\begin{bmatrix} -R\ddot{\psi}\sin\psi\sin\vartheta + R\ddot{\vartheta}\cos\psi\cos\vartheta + \ddot{R}\cos\psi\sin\vartheta - R\dot{\psi}^2\cos\psi\sin\vartheta \\ -2R\dot{\psi}\dot{\vartheta}\sin\psi\cos\vartheta - 2\dot{R}\dot{\psi}\sin\psi\sin\vartheta - R\dot{\vartheta}^2\cos\psi\sin\vartheta + 2\dot{R}\dot{\vartheta}\cos\psi\cos\vartheta \\ \\ R\ddot{\psi}\cos\psi\sin\vartheta + R\ddot{\vartheta}\sin\psi\cos\vartheta + \ddot{R}\sin\psi\sin\vartheta - R\dot{\psi}^2\sin\psi\sin\vartheta \\ +2R\dot{\psi}\dot{\vartheta}\cos\psi\cos\vartheta + 2\dot{R}\dot{\psi}\cos\psi\sin\vartheta - R\dot{\vartheta}^2\sin\psi\sin\vartheta + 2\dot{R}\dot{\vartheta}\sin\psi\cos\vartheta \\ \\ -R\ddot{\vartheta}\sin\vartheta + \ddot{R}\cos\vartheta - R\dot{\vartheta}^2\cos\vartheta - 2\dot{R}\dot{\vartheta}\sin\vartheta \end{bmatrix}. \tag{2.18}$$

The acceleration vector is linearly dependent on the second derivatives and quadratically dependent on the first derivatives of the generalized coordinates.

End of Example 2.1.

By introducing generalized coordinates, the uniqueness of the kinematic description at singular points may be lost due to a loss of degrees of freedoms. Thus we should always require the full rank of the Jacobian matrix or

$$\det \boldsymbol{H}_T \neq 0. \tag{2.19}$$

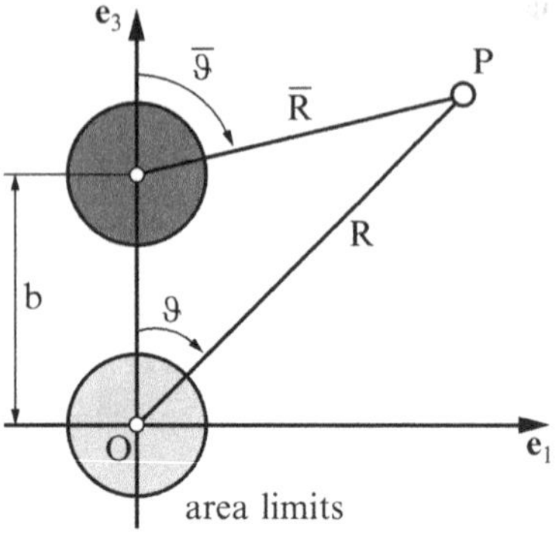

Fig. 2.3 Definition of complementary spherical coordinates

In Example 2.1, according to (2.16) there is a rank decrease of two in matrix $\boldsymbol{H}_T$ for $R = 0$, which certainly violates (2.19). The explanation for this is that the point P for $R \to 0$ can now only move in the direction

$$\boldsymbol{e}_R = \begin{bmatrix} \cos\psi\sin\vartheta & \sin\psi\sin\vartheta & \cos\vartheta \end{bmatrix}, \tag{2.20}$$

so it now only has one degree of freedom. This problem can be solved by the introducing complementary spherical coordinates $\overline{\psi}$, $\overline{\vartheta}$, see Fig. 2.3. By extending (2.15), we then obtain

$$\boldsymbol{r}(t) = \begin{bmatrix} R\cos\psi\sin\vartheta \\ R\sin\psi\sin\vartheta \\ R\cos\vartheta \end{bmatrix} = \begin{bmatrix} \overline{R}\cos\psi\sin\overline{\vartheta} \\ \overline{R}\sin\psi\sin\overline{\vartheta} \\ \overline{R}\cos\overline{\vartheta} + b \end{bmatrix}, \tag{2.21}$$

i.e. two different singular points $R = 0$ and $\overline{R} = 0$ appear, where $b > 0$ is an arbitrary distance. If we now limit, for example, the critical generalized coordinates by the areas shown in Fig. 2.3,

$$R \geq b/4, \qquad \overline{R} \geq b/4, \tag{2.22}$$

then with position vectors that are complementary to each other $\boldsymbol{x}(t)$ and $\overline{\boldsymbol{x}}(t)$, a unique description of position is always possible. If one of the limits (2.22) is violated, transition to the complementary spherical coordinates takes place and vice versa. Following (2.21), we have for this, for example, the following relation

$$\overline{R} = R\frac{\sin\vartheta}{\sin\overline{\vartheta}}, \qquad \cot\overline{\vartheta} = \cot\vartheta - \frac{b}{R\sin\vartheta}. \tag{2.23}$$

For many motions, the singular points are not critical. For example, the planets always move at a great distance from the singular point at the origin. Yet in the case of rotating rigid bodies we constantly find singular points, to which numerous

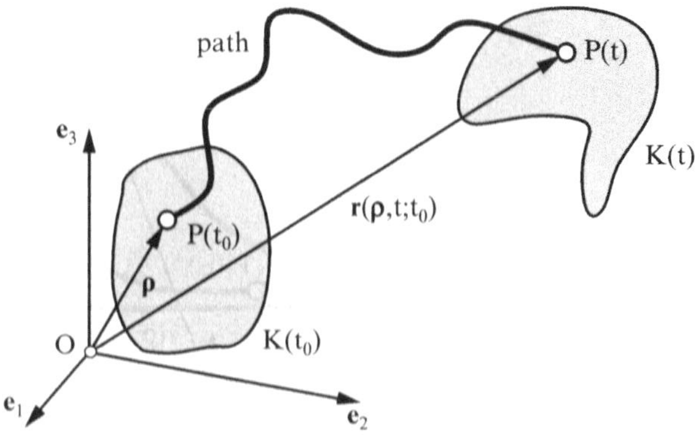

Fig. 2.4 Motion of a free body

papers have been dedicated in gyroscope theory. It is therefore worthwhile to deal with this problem already at the level of the point motion.

A free system of p mass points in space has $3p$ degrees of freedom. If we merge the $3p$ generalized coordinates of the total system into a $3p \times 1$ position vector $\boldsymbol{x}(t)$, the ith point is defined in accordance with (2.4) as

$$\boldsymbol{r}_i(t) = \boldsymbol{r}_i(\boldsymbol{x}), \qquad i = 1(1)p. \tag{2.24}$$

The relations (2.5), (2.6) and (2.12), (2.13) are also true for mass point systems. In particular, (2.8) turns into a $3 \times 3p$ Jacobian matrix $\boldsymbol{H}_{Ti}(\boldsymbol{x}), i = 1(1)p$.

2.1.2 Kinematics of a Rigid Body

The rigid body is a simple model of continuum mechanics. Like all continua, it consists of a coherent, compact quantity of mass points. However, in a rigid body, the distances between arbitrary mass points are constant. From the standpoint of continuum mechanics, a rigid body is thus free of strain. Yet it is also statically indeterminate, i.e. the forces and stresses arising in its interior cannot be calculated, see Sect. 5.4.2. Nonetheless, the rigid body is eminently suitable for the investigation of motions in many contexts within dynamics. This is especially the case for systems of rigid bodies, called multibody systems.

In order to describe free multibody systems kinematically, such as they appear in rotor dynamics for example, it is again sufficient to consider a single rigid body. In a free system, all rigid bodies are kinematically equal.

An arbitrary, rigid or nonrigid body K is described mathematically by its reference configuration, i.e. a constant and reversibly unique assignment of location vectors $\boldsymbol{\rho}$ to the mass points, see Fig. 2.4. If nothing else is stipulated, we use an

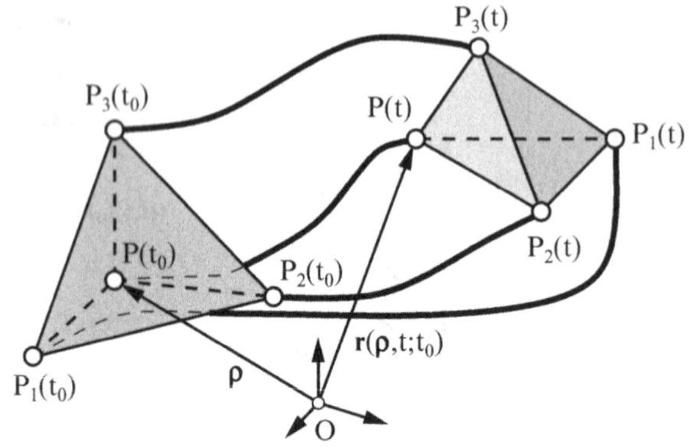

Fig. 2.5 Motion of a free tetrahedral element

inertial Cartesian frame $\{O,\boldsymbol{e}_\alpha\},\alpha = 1(1)3$, and a nonpolar continuum. The current configuration of a body $K(t)$ in motion at time t in space,

$$\boldsymbol{r} = \boldsymbol{r}(\boldsymbol{\rho},t;t_0), \tag{2.25}$$

is referred to the reference configuration of the body $K(t_0)$ at the reference time t_0,

$$\boldsymbol{\rho} = \boldsymbol{r}(\boldsymbol{\rho},t_0;t_0). \tag{2.26}$$

On the other hand, the location vectors $\boldsymbol{\rho}$ are also determined by the inverse function of (2.25),

$$\boldsymbol{\rho} = \boldsymbol{\rho}(\boldsymbol{r},t;t_0), \tag{2.27}$$

yielding a unique assignment to the mass points. The fixed reference time t_0 is taken as the basis for Eqs. (2.25)–(2.27). Yet it is also possible to select the running time $t_0 = t$ as the reference time. We then obtain $\boldsymbol{\rho}(\boldsymbol{r},t;t) = \boldsymbol{r}$, i.e. the mass point P designated by $\boldsymbol{\rho}$ coincides at the moment with the point in space described by $\boldsymbol{r}$. The running reference time t will prove useful for determining the current rotation velocity vector.

In the following, the variables $\boldsymbol{\rho}$, t and t_0 will only be written if needed. This does not affect the explicit dependence of the parameters considered in these variables. The coordinates of the vector $\boldsymbol{\rho}$ are also called material coordinates, while the coordinates of the vector $\boldsymbol{r}$ are designated as spatial coordinates.

The general motion of a nonrigid body K is composed of rotations and strains. This motion is called deformation. Since deformation changes from point to point within the body, it is properly characterized by the deformation gradient $\boldsymbol{F}(\boldsymbol{\rho},t;t_0) = \partial\boldsymbol{r}/\partial\boldsymbol{\rho}$. The deformation gradient describes, for example, the motion of a tetrahedral element from a reference configuration into the current configuration, as shown in Fig. 2.5. A tetrahedral element comprises four infinitesimally neighboring mass points P, P_1, P_2, P_3. The line elements between point P and points P_1,P_2,P_3 are thus transformed from the respective reference configuration $d\boldsymbol{\rho}$ into the respective current configuration $d\boldsymbol{r}$. This transformation is effected by the deformation gradient $\boldsymbol{F}(\boldsymbol{\rho},t;t_0)$. From (2.25) we obtain

$$\boldsymbol{r}(\boldsymbol{\rho}+d\boldsymbol{\rho},t;t_0)-\boldsymbol{r}(\boldsymbol{\rho},t;t_0)=d\boldsymbol{r}=\frac{\partial \boldsymbol{r}}{\partial \boldsymbol{\rho}}\cdot d\boldsymbol{\rho}=\boldsymbol{F}(\boldsymbol{\rho},t;t_0)\cdot d\boldsymbol{\rho}. \tag{2.28}$$

Conversely, we obtain with (2.27) and (2.25)

$$d\boldsymbol{\rho}=\frac{\partial \boldsymbol{\rho}}{\partial \boldsymbol{r}}\cdot d\boldsymbol{r}=\boldsymbol{F}^{-1}(\boldsymbol{\rho},t;t_0)\cdot d\boldsymbol{r}. \tag{2.29}$$

In order to avoid ambiguity in (2.25) and (2.27), the deformation gradient in (2.28) and (2.29) must be always regular, $\det\boldsymbol{F}\neq 0$. Due to (2.26), (2.28) also results in $\boldsymbol{F}(\boldsymbol{\rho},t_0;t_0)=\boldsymbol{E}$ and thus $\det\boldsymbol{F}(\boldsymbol{\rho},t_0;t_0)=+1$. Here $\boldsymbol{E}$ is again the 3×3 unit tensor.

Furthermore, if we take the consistency of the deformation into consideration, we obtain the condition

$$\det\boldsymbol{F}>0. \tag{2.30}$$

The current configuration of the tetrahedral element under consideration is determined by a total of 12 coordinates corresponding to the 12 degrees of freedom of the 4 mass points. These 12 coordinates can also be interpreted as the 3 displacement coordinates of the 3×1 location vector at point P and the 9 coordinates of the 3×3 tensor of the deformation gradient.

For a rigid body K, the pairwise distances of all mass points remain constant during deformation,

$$d\boldsymbol{r}\cdot d\boldsymbol{r}=\boldsymbol{F}\cdot d\boldsymbol{\rho}\cdot\boldsymbol{F}\cdot d\boldsymbol{\rho}=d\boldsymbol{\rho}\cdot\boldsymbol{F}^T\cdot\boldsymbol{F}\cdot d\boldsymbol{\rho}\stackrel{!}{=}d\boldsymbol{\rho}\cdot d\boldsymbol{\rho}. \tag{2.31}$$

We thus find for the deformation gradient of the rigid body

$$\boldsymbol{F}^T\cdot\boldsymbol{F}=\boldsymbol{E}. \tag{2.32}$$

Thus this deformation gradient $\boldsymbol{F}$ is independent of the location vector $\boldsymbol{\rho}$ of the mass points of the rigid body. It can therefore only be a function of time t. The deformation gradient thus corresponds to the 3×3 rotation tensor $\boldsymbol{S}(t;t_0)$ of the rigid body,

$$\boldsymbol{F}(\boldsymbol{\rho},t;t_0)=\boldsymbol{S}(t;t_0). \tag{2.33}$$

As a result of (2.30) and (2.32), the rotation tensor $\boldsymbol{S}(t;t_0)$ is actually an orthogonal tensor. In the following, the rotation tensor will always be applied to the reference configuration, so we need not write the reference time t_0.

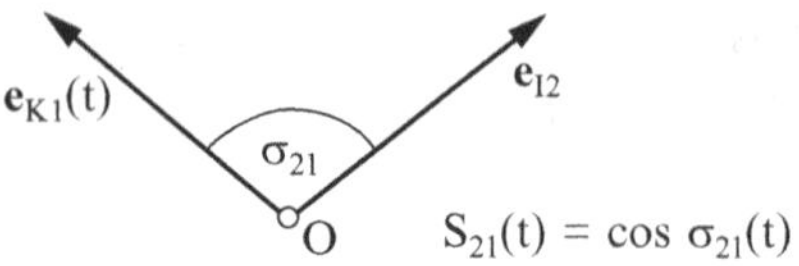

Fig. 2.6 Direction cosine

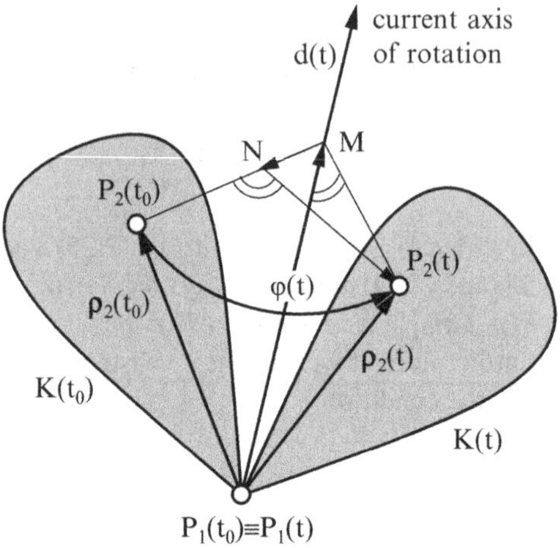

Fig. 2.7 Finite rotation of a rigid body

We will now introduce the properties of the rotation of a rigid body in detail. Various possible methods of description will be used in the process, either via the nine direction cosines, four rotation parameters or three rotation angles. In each case, there remain three generalized coordinates corresponding to the three degrees of freedom of the rotation of a rigid body.

Every Cartesian coordinate $S_{\alpha\beta}(t)$, $\alpha,\beta = 1(1)3$, of the rotation tensor (2.33) can be viewed as a direction cosine of the angle $\sigma_{\alpha\beta}(t)$ between the basis vector $\boldsymbol{e}_{I\alpha}$ of the spatially fixed inertial frame I and the basis vector $\boldsymbol{e}_{K\beta}(t)$ of the corresponding body-fixed frame K, see Fig. 2.6. The Cartesian body-fixed frame $\{P(t);\boldsymbol{e}_{K\beta}(t)\}$, $\beta = 1(1)3$ here coincides at time $t = t_0$ with the inertial frame,

$$\{P(t_0);\boldsymbol{e}_{K\beta}(t_0)\} = \{0;\boldsymbol{e}_{I\alpha}\}. \tag{2.34}$$

The nine direction cosines $S_{\alpha\beta}, \alpha, \beta = 1(1)3$, are subject to the six orthogonality constraints (2.32), so only three generalized coordinates remain.

The rotation tensor $\boldsymbol{S}$ according to (2.33) can also be expressed by the four rotation parameters, i.e. the three coordinates of the vector $\boldsymbol{d}$, normalized to length one, of the rotation axis and the scalar rotation angle $\varphi(t)$. The representation of a finite rotation by its rotation axis and a rotation angle is due to Euler. For this reason, we also denote the four rotation parameters as Euler parameters.

From Fig. 2.7 we conclude on the one hand

$$\boldsymbol{\rho}_2(t) = \boldsymbol{S}(t) \cdot \boldsymbol{\rho}_2(t_0), \tag{2.35}$$

while on the other hand we obtain from the vector polygon P_1MNP_2 the relation

$$\boldsymbol{\rho}_2(t) = \boldsymbol{dd}\cdot\boldsymbol{\rho}_2(t_0) + (\boldsymbol{\rho}_2(t_0) - \boldsymbol{dd}\cdot\boldsymbol{\rho}_2(t_0))\cos\varphi + \tilde{\boldsymbol{d}}\cdot\boldsymbol{\rho}_2(t_0)\sin\varphi. \tag{2.36}$$

Comparison of (2.35) and (2.36) directly yields

$$\boldsymbol{S}(t) = \boldsymbol{dd} + (\boldsymbol{E} - \boldsymbol{dd})\cos\varphi + \tilde{\boldsymbol{d}}\sin\varphi. \tag{2.37}$$

The skew-symmetric 3×3 tensor $\tilde{\boldsymbol{d}}$ of the 3×1 vector $\boldsymbol{d}$ and its dyadic product $\boldsymbol{dd}$ were introduced into (2.36), see Appendix A,

$$\boldsymbol{d} = \begin{bmatrix} d_1 \\ d_2 \\ d_3 \end{bmatrix}, \quad \tilde{\boldsymbol{d}} = -\tilde{\boldsymbol{d}}^T = \begin{bmatrix} 0 & -d_3 & d_2 \\ d_3 & 0 & -d_1 \\ -d_2 & d_1 & 0 \end{bmatrix}, \quad \boldsymbol{dd} = \begin{bmatrix} d_1d_1 & d_1d_2 & d_1d_3 \\ d_2d_1 & d_2d_2 & d_2d_3 \\ d_3d_1 & d_3d_2 & d_3d_3 \end{bmatrix}. \tag{2.38}$$

The skew-symmetric tensor of a vector is denoted by the symbol ($\tilde{\ }$). It gives the cross or vector product

$$\tilde{\boldsymbol{a}}\cdot\boldsymbol{b} = \boldsymbol{a}\times\boldsymbol{b}. \tag{2.39}$$

Between the dyadic product $\boldsymbol{ab}$, the scalar product $\boldsymbol{a}\cdot\boldsymbol{b} = \boldsymbol{b}\cdot\boldsymbol{a}$, and the expanded vector product $\tilde{\boldsymbol{a}}\cdot\tilde{\boldsymbol{b}}$, there is also, using (A.30), the useful relation

$$\boldsymbol{ab} = (\boldsymbol{b}\cdot\boldsymbol{a})\boldsymbol{E} + \tilde{\boldsymbol{b}}\cdot\tilde{\boldsymbol{a}}. \tag{2.40}$$

If we now consider that the vector $\boldsymbol{d}$ of the rotation axis is a unit vector,

$$\boldsymbol{d}\cdot\boldsymbol{d} = 1, \tag{2.41}$$

which corresponds exactly to a constraint between the four rotation parameters $d_\alpha, \alpha = 1(1)3$, and φ, we then obtain from (2.40) the relation

$$\boldsymbol{dd} = \boldsymbol{E} + \tilde{\boldsymbol{d}}\cdot\tilde{\boldsymbol{d}}. \tag{2.42}$$

Then (2.37) can be written as

$$\boldsymbol{S}(t) = \boldsymbol{E} + \tilde{\boldsymbol{d}}\sin\varphi + \tilde{\boldsymbol{d}}\cdot\tilde{\boldsymbol{d}}(1-\cos\varphi). \tag{2.43}$$

We can see that for $t = t_0$ the rotation tensor turns into the unit tensor $\boldsymbol{E}$ due to $\varphi(t_0) = 0$.

Closely related to the four rotation parameters are the four quaternions $q_n(t)$, $n = 0(1)3$, which we obtain following the transition to half the rotation angle

$$q_0 = \cos\frac{\varphi}{2}, \qquad \boldsymbol{q} = \begin{bmatrix} q_1 \\ q_2 \\ q_3 \end{bmatrix} = \boldsymbol{d}\sin\frac{\varphi}{2}. \tag{2.44}$$

Equation (2.43) thereby acquires the form

$$S(t) = E + 2q_0\tilde{q} + 2\tilde{q} \cdot \tilde{q} \tag{2.45}$$

and the constraint (2.41) becomes

$$q_0^2 + q \cdot q = 1. \tag{2.46}$$

The three Rodrigues parameters $p_\alpha(t), \alpha = 1(1)3$, are obtained by normalizing the quaternions. They can be represented as a 3×1 vector $\boldsymbol{p}$,

$$p = d \tan\frac{\varphi}{2} = \frac{1}{q_0} q. \tag{2.47}$$

The rotation tensor then has the form

$$S(t) = E + 2\frac{\tilde{p} + \tilde{p} \cdot \tilde{p}}{1 + p \cdot p} \tag{2.48}$$

which is a nice, compact expression.

The four rotation parameters can also be determined conversely from the rotation tensor. To achieve this we can utilize, for example, the fact that a truly orthogonal 3×3 tensor has the eigenvalues $\lambda_1 = 1$, $\lambda_{2,3} = e^{\pm i\varphi}$. The eigenvector belonging to the real eigenvalue describes the rotation axis, while the argument φ of the imaginary eigenvalues gives the rotation angle. However, the rotation direction cannot be found by solving the eigenvalue problem. To do this, an additional comparison with the rotation tensor (2.43) is required. For $\varphi = 0, 2\pi, 4\pi, \ldots$, the rotation tensor has a triple eigenvalue $\lambda_{1,2,3} = 1$. Then each unit vector is also an eigenvector and consequently also a rotation axis.

Example 2.2 (Rotation Axis and Rotation Angle of a Rigid Body). Let a rotation tensor $S(t)$ be given by

$$S(t) = \begin{bmatrix} \cos\vartheta & 0 & -\sin\vartheta \\ 0 & 1 & 0 \\ \sin\vartheta & 0 & \cos\vartheta \end{bmatrix}. \tag{2.49}$$

The eigenvalue problem

$$(\lambda E - S) \cdot d = 0 \tag{2.50}$$

provides the characteristic equation

$$(\lambda - 1)(\lambda^2 - 2\lambda\cos\vartheta + 1) = 0 \tag{2.51}$$

with the eigenvalues

$$\lambda_1 = 1, \qquad \lambda_{2,3} = e^{\pm i\vartheta}. \tag{2.52}$$

The first normalized eigenvector is

$$\boldsymbol{d} = \begin{bmatrix} 0 & -1 & 0 \end{bmatrix}, \tag{2.53}$$

where the sign has been determined by inserting into (2.43) and comparing with (2.49). For the quaternions we obtain

$$\begin{aligned} q_0^2(t) &= \frac{1}{2}(1+\cos\vartheta) = \cos^2\frac{\vartheta}{2}, & q_1^2(t) &= 0, \\ q_2^2(t) &= \frac{1}{2}(1-\cos\vartheta) = \sin^2\frac{\vartheta}{2}, & q_3^2(t) &= 0. \end{aligned} \tag{2.54}$$

As a result of the quadratic quantities, here also the rotation direction must be determined by comparison with (2.49).

End of Example 2.2.

The four rotation parameters $d_\alpha(t), \alpha = 1(1)3$, and $\varphi(t)$ and the four quaternions $q_n(t), n = 0(1)3$, are subject to exactly one constraint, so here again only three generalized coordinates remain. The three Rodrigues parameters $p_\alpha(t)$ from (2.47) on the other hand can be used directly as generalized coordinates. However, their engineering applicability is limited by the infinite values of the tangent function for $\varphi = \pi/2, 3\pi/2, 5\pi/2, \ldots$.

Finally, the rotation tensor (2.33) can also be expressed by means of three rotation angles from elementary rotations. Elementary rotations exist when the rotation axis coincides with one of the coordinate axes. They are defined by the name of the rotation angle and the specification of the rotation axis. There are three elementary rotation matrices, these corresponding to the three basis vectors of a Cartesian frame.

In order to construct a unique rotation tensor, we now make use of the property that orthogonality is preserved in the multiplication of orthogonal tensors, and we restrict ourselves also to three independent angles as generalized coordinates

$$\boldsymbol{\alpha}_1(t) = \begin{bmatrix} 1 & 0 & 0 \\ 0 & \cos\alpha & -\sin\alpha \\ 0 & \sin\alpha & \cos\alpha \end{bmatrix}, \tag{2.55}$$

$$\boldsymbol{\beta}_2(t) = \begin{bmatrix} \cos\beta & 0 & \sin\beta \\ 0 & 1 & 0 \\ -\sin\beta & 0 & \cos\beta \end{bmatrix}, \tag{2.56}$$

$$\boldsymbol{\gamma}_3(t) = \begin{bmatrix} \cos\gamma & -\sin\gamma & 0 \\ \sin\gamma & \cos\gamma & 0 \\ 0 & 0 & 1 \end{bmatrix}. \tag{2.57}$$

Among the numerous possible ways to describe finite rotations with three generalized coordinates, here we will mention the Euler angles

$$\boldsymbol{S}(t) = \boldsymbol{\psi}_3(t) \cdot \boldsymbol{\vartheta}_1(t) \cdot \boldsymbol{\varphi}_3(t) \tag{2.58}$$

and the Cardano angles

$$\boldsymbol{S}(t) = \boldsymbol{\alpha}_1(t) \cdot \boldsymbol{\beta}_2(t) \cdot \boldsymbol{\gamma}_3(t). \tag{2.59}$$

In the construction of rotation tensors from elementary rotations, it should also be kept in mind that the tensor product is not commutative. For this reason, a complete definition includes not only the angle name and rotation axis, but also the sequence of elementary rotations. If we now evaluate (2.59) with (2.55)–(2.57), we obtain

$$\boldsymbol{S}(t) = \begin{bmatrix} \cos\beta\cos\gamma & -\cos\beta\sin\gamma & \sin\beta \\ \begin{matrix}\cos\alpha\sin\gamma \\ +\sin\alpha\sin\beta\cos\gamma\end{matrix} & \begin{matrix}\cos\alpha\cos\gamma \\ -\sin\alpha\sin\beta\sin\gamma\end{matrix} & -\sin\alpha\cos\beta \\ \begin{matrix}\sin\alpha\sin\gamma \\ -\cos\alpha\sin\beta\cos\gamma\end{matrix} & \begin{matrix}\sin\alpha\cos\gamma \\ +\cos\alpha\sin\beta\sin\gamma\end{matrix} & \cos\alpha\cos\beta \end{bmatrix}. \tag{2.60}$$

The Cardano angles can now be found conversely from the rotation tensor. For this purpose it is appropriate to use the sparsely populated coordinates, e.g.,

$$\sin\beta = S_{13}, \qquad \cos\alpha = \frac{S_{33}}{\cos\beta}, \qquad \cos\gamma = \frac{S_{11}}{\cos\beta}. \tag{2.61}$$

Singular rotation angles $\beta = \pi/2, 3\pi/2, 5\pi/2, \ldots$ exist here for $\cos\beta = 0$. They come from the fact that two elementary rotation axes coincide and thus one degree of freedom in the rotation is lost. This is especially clear if we consider, for example, the rotation tensor (2.60) in the area of a singularity, $\alpha = \Delta\alpha$, $\beta = \pi/2 + \Delta\beta$, $\gamma = \Delta\gamma$ with $\Delta\alpha, \Delta\beta, \Delta\gamma \ll 1$,

$$\boldsymbol{S}(t) = \begin{bmatrix} -\Delta\beta & 0 & 1 \\ (\Delta\alpha + \Delta\gamma) & 1 & 0 \\ -1 & (\Delta\alpha + \Delta\gamma) & -\Delta\beta \end{bmatrix}. \tag{2.62}$$

Only the sum of the angles $(\Delta\alpha + \Delta\gamma)$ and the single angle $\Delta\beta$ remain as generalized coordinates.

The singularities of rotation angles can be avoided by limiting the angle of the second elementary rotation and by introducing complementary rotation angles. If we limit, e.g., the second Cardano angle

$$-\pi/3 < \beta < \pi/3, \tag{2.63}$$

and if we add the complementary Cardano angle to (2.59)

$$\boldsymbol{S}(t) = \overline{\boldsymbol{\alpha}}_1(t) \cdot \overline{\boldsymbol{\beta}}_2(t) \cdot \overline{\boldsymbol{\gamma}}_1(t), \qquad |\overline{\beta}| > \pi/6, \tag{2.64}$$

then no singularity arises. For $\overline{\alpha} = \Delta\overline{\alpha}$, $\overline{\beta} = \frac{\pi}{2} + \Delta\overline{\beta}$, $\overline{\gamma} = \Delta\overline{\gamma}$ with $\Delta\overline{\alpha}$, $\Delta\overline{\beta}$, $\Delta\overline{\gamma} \ll 1$ we obtain from (2.64) the rotation tensor

$$\boldsymbol{S}(t) = \begin{bmatrix} -\Delta\overline{\beta} & \Delta\overline{\gamma} & 1 \\ \Delta\overline{\alpha} & 1 & \Delta\overline{\gamma} \\ -1 & \Delta\overline{\alpha} & -\Delta\overline{\beta} \end{bmatrix}. \tag{2.65}$$

With this we obtain three independent coordinates $\Delta\overline{\alpha}, \Delta\overline{\beta}, \Delta\overline{\gamma}$. At the boundaries (2.63) and (2.64), transformation of the angle takes place over the sparsely populated coordinates of both rotation tensors. The intersecting boundaries guarantee however a low number of shifts between (2.59) and (2.64). In particular, the relations

$$\beta = \arcsin(\sin\overline{\beta}\cos\overline{\gamma}), \tag{2.66}$$

$$\sin\alpha = \frac{1}{\cos\beta}(\cos\overline{\alpha}\sin\overline{\gamma} + \sin\overline{\alpha}\cos\overline{\beta}\cos\overline{\gamma}), \tag{2.67}$$

$$\cos\alpha = \frac{1}{\cos\beta}(-\sin\overline{\alpha}\sin\overline{\gamma} + \cos\overline{\alpha}\cos\overline{\beta}\cos\overline{\gamma}), \tag{2.68}$$

$$\sin\gamma = -\frac{1}{\cos\beta}(\sin\overline{\beta}\sin\overline{\gamma}), \tag{2.69}$$

$$\cos\gamma = \frac{1}{\cos\beta}\cos\overline{\beta} \tag{2.70}$$

apply and the complementary relations

$$\overline{\beta} = \arccos(\cos\beta\cos\gamma), \tag{2.71}$$

$$\sin\overline{\alpha} = \frac{1}{\sin\overline{\beta}}(\cos\alpha\sin\gamma + \sin\alpha\sin\beta\cos\gamma), \tag{2.72}$$

$$\cos\overline{\alpha} = \frac{1}{\sin\overline{\beta}}(-\sin\alpha\sin\gamma + \cos\alpha\sin\beta\cos\gamma), \tag{2.73}$$

$$\sin\overline{\gamma} = -\frac{1}{\sin\overline{\beta}}(\cos\beta\sin\gamma), \tag{2.74}$$

$$\cos\overline{\gamma} = \frac{1}{\sin\overline{\beta}}\sin\beta. \tag{2.75}$$

The elementary rotations permit us to construct, by means of numerous combination possibilities, a suitable rotation tensor for every engineering problem. Especially

Table 2.1 Possible methods for describing the rotation of a rigid body

Coordinates of the rotation tensor	Relations of the coordinates	Generalized coordinates
9 direction cosines	6 constraints	e.g.
$\boldsymbol{S}(t)$	$\boldsymbol{S}\cdot\boldsymbol{S}^T=\boldsymbol{E}$	$S_{11}(t),S_{12}(t),S_{23}(t)$
4 rotation parameters	1 constraint	e.g.
$\boldsymbol{d}(t),\varphi(t)$	$\boldsymbol{d}\cdot\boldsymbol{d}=1$	$d_1(t),d_2(t),\varphi(t)$
4 quaternions	1 constraint	e.g.
$q_0(t)\ \boldsymbol{q}(t)$	$q_0^2+\boldsymbol{q}\cdot\boldsymbol{q}=1$	$q_0(t),q_1(t),q_2(t)$
3 Euler angles		
$\psi(t),\vartheta(t),\varphi(t)$	–	$\psi(t),\vartheta(t),\varphi(t)$
3 Cardano angles		
$\alpha(t),\beta(t),\gamma(t)$	–	$\alpha(t),\beta(t),\gamma(t)$

flight mechanics and theory of gyroscopes make extensive use of this, see e.g. Magnus [35] or Arnold and Maunder [2].

Possible methods of describing the rotation of a rigid body are summarized in Table 2.1. If we merge the remaining generalized coordinates of rotation again in a 3×1 position vector

$$\boldsymbol{x}(t)=\begin{bmatrix}x_1\ x_2\ x_3\end{bmatrix} \tag{2.76}$$

then the following applies quite generally,

$$\boldsymbol{S}(t)=\boldsymbol{S}(\boldsymbol{x}(t))=\boldsymbol{S}(\boldsymbol{x}), \tag{2.77}$$

independent of the particular choice of generalized coordinates.

A rigid body is in a state of general motion when its rotation is complemented with displacement. According to Fig. 2.8, the current configuration of the rigid body K is then

$$\boldsymbol{r}(\boldsymbol{\rho},t)=\boldsymbol{r}_1(t)+\boldsymbol{r}_P(\boldsymbol{\rho},t)=\boldsymbol{r}_1(t)+\boldsymbol{S}(t)\cdot\boldsymbol{\rho}. \tag{2.78}$$

This equation, first introduced in an intuitive manner, can also be found formally by integrating (2.28) with a fixed reference time t_0. In the case of the rigid body, this integration is possible in a closed form since the deformation gradient, in accordance with (2.33), does not depend on the mass coordinates.

In (2.78), $\boldsymbol{r}_1(t)$ is the 3×1 location vector of point $P_1(t)$. It describes the translation of the rigid body. The 3×3 rotation tensor $\boldsymbol{S}(t)$ denotes the rotation of the rigid body. Translation, as opposed to rotation, does not contribute to the deformation gradient, as follows from (2.78), (2.28). Furthermore, the following applies according to (2.29), (2.32) and (2.33) for the inverse deformation,

$$\boldsymbol{\rho}=\boldsymbol{S}^T(t)\cdot\boldsymbol{r}_P(\boldsymbol{\rho},t). \tag{2.79}$$

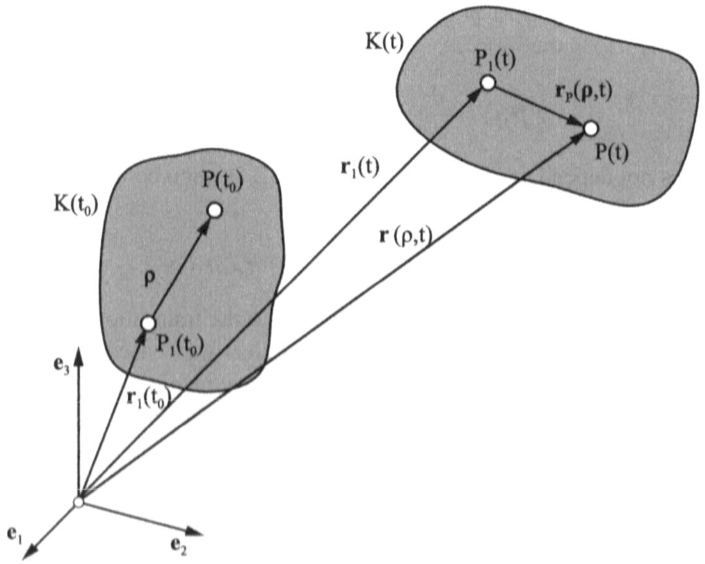

Fig. 2.8 Motion of a free rigid body

Inserted into (2.78) we obtain

$$
\begin{aligned}
\boldsymbol{r}(\boldsymbol{\rho},t) &= \boldsymbol{r}_1(t) + \boldsymbol{S}(t;t_0)\cdot\boldsymbol{S}^T(t;t_0)\cdot\boldsymbol{r}_P(\boldsymbol{\rho},t) \\
&= \boldsymbol{r}_1(t) + \boldsymbol{S}(t;t_0)\cdot\boldsymbol{S}(t_0;t)\cdot\boldsymbol{r}_P(\boldsymbol{\rho},t) = \boldsymbol{r}_1(t) + \boldsymbol{S}(t;t)\cdot\boldsymbol{r}_P(\boldsymbol{\rho},t) \\
&= \boldsymbol{r}_1(t) + \boldsymbol{r}_P(\boldsymbol{\rho},t).
\end{aligned} \tag{2.80}
$$

The instantaneous rotation tensor $\boldsymbol{S}(t,t) = \boldsymbol{E}$ is independent of the reference time t_0. It thus has the property of being a field in the sense of continuum mechanics.

A free rigid body has six degrees of freedom. For the three degrees of freedom of translation all the relations of point kinematics are valid, see Sect. 2.1.1. According to Table 2.1, the three degrees of freedom of rotation also require three generalized coordinates, so on the whole the 6×1 position vector

$$\boldsymbol{x}(t) = \begin{bmatrix} x_1 & x_2 & x_3 & x_4 & x_5 & x_6 \end{bmatrix} \tag{2.81}$$

describes the general motion of a rigid body. The location vector and rotation tensor of the rigid body are thus written

$$\boldsymbol{r}(t) = \boldsymbol{r}(\boldsymbol{x}), \qquad \boldsymbol{S}(t) = \boldsymbol{S}(\boldsymbol{x}). \tag{2.82}$$

In special cases of pure translation or rotation (2.82) becomes (2.4) or (2.77), whereby the number of degrees of freedom is in both cases reduced to three.

The current velocity of a point on the rigid body K is obtained by the time derivative of (2.78) in the form

$$\boldsymbol{v}(\boldsymbol{\rho},t) = \frac{d}{dt}\boldsymbol{r}(\boldsymbol{\rho},t) = \dot{\boldsymbol{r}}_1(t) + \dot{\boldsymbol{S}}(t)\cdot\boldsymbol{\rho} \tag{2.83}$$

since $\boldsymbol{\rho}$ does not depend on time. If we also take (2.79) into consideration, we can also write

$$\boldsymbol{v}(\boldsymbol{\rho},t) = \dot{\boldsymbol{r}}_1(t) + \dot{\boldsymbol{S}}(t)\cdot\boldsymbol{S}^T(t)\cdot\boldsymbol{r}_P(\boldsymbol{\rho},t). \tag{2.84}$$

The first term on the right-hand side corresponds to the translation velocity of the reference point P_1 from (2.5). The second term clearly is based on rotation and should be inspected more closely here. From (2.84) we obtain, with a Taylor series expansion with respect to dt taking into account the orthogonality of $\boldsymbol{S}$, the result

$$\begin{aligned}\dot{\boldsymbol{S}}(t)\cdot\boldsymbol{S}^T(t) &= \frac{\boldsymbol{S}(t+dt;t_0)-\boldsymbol{S}(t;t_0)}{dt}\cdot\boldsymbol{S}^T(t;t_0)\\ &= \frac{\boldsymbol{S}(t+dt;t)-\boldsymbol{E}}{dt} = \tilde{\boldsymbol{d}}(t;t)\frac{d\varphi(t;t)}{dt} = \frac{d\tilde{s}(t)}{dt} = \tilde{\omega}(t).\end{aligned} \tag{2.85}$$

Here $\tilde{\boldsymbol{d}}(t;t)$ denotes the rotation axis and $\dot{\varphi}(t;t)$ the velocity of the instantaneous rotation. Also, $d\tilde{s}(t) = \tilde{\boldsymbol{d}}(t;t)d\varphi(t;t)$ is designated as the 3×3 tensor of infinitesimal instantaneous rotation and $\tilde{\boldsymbol{\omega}}(t)$ as the 3×3 tensor of rotation velocity. In accordance with (2.38), to this tensor is assigned the 3×1 vector $\boldsymbol{\omega}(t)$ of rotation velocity.

Infinitesimal rotation $ds(t)$ thus has, in contrast to finite rotation, the property of being a vector. Moreover, it also no longer depends on the reference time and is thus a field quantity in the sense of continuum mechanics. Accordingly, (2.84) can also be written as

$$\boldsymbol{v}(\boldsymbol{\rho},t) = \boldsymbol{v}_1(t) + \tilde{\boldsymbol{\omega}}(t)\cdot\boldsymbol{r}_P(\boldsymbol{\rho},t) = \boldsymbol{v}_1(t) + \boldsymbol{\omega}(t)\times\boldsymbol{r}_P(\boldsymbol{\rho},t), \tag{2.86}$$

which corresponds to the known formula for the velocity field of a rigid body. The 3×1 vectors $\boldsymbol{v}_1(t)$ and $\boldsymbol{\omega}(t)$ clearly describe the state of velocity of the rigid body. They can also be merged into the 6×1 twist $(\boldsymbol{v}_1(t),\boldsymbol{\omega}(t))$, see Sect. 5.7.2.

In order to calculate the rotational velocity vector $\boldsymbol{\omega}(t)$, we thus have relation (2.85) available which means a formal time differentiation of the rotation tensor.

Example 2.3 (Rotational Velocity of a Rigid Body). With the rotation tensor (2.49) from Example 2.2 we get the rotational velocity tensor

$$\begin{aligned}\tilde{\boldsymbol{\omega}}(t) &= \dot{\vartheta}\begin{bmatrix}-\sin\vartheta & 0 & -\cos\vartheta\\ 0 & 0 & 0\\ \cos\vartheta & 0 & -\sin\vartheta\end{bmatrix}\cdot\begin{bmatrix}\cos\vartheta & 0 & \sin\vartheta\\ 0 & 1 & 0\\ -\sin\vartheta & 0 & \cos\vartheta\end{bmatrix}\\ &= \dot{\vartheta}\begin{bmatrix}0 & 0 & -1\\ 0 & 0 & 0\\ 1 & 0 & 0\end{bmatrix}\end{aligned} \tag{2.87}$$

and thereby the rotation velocity vector

$$\boldsymbol{\omega}(t) = \begin{bmatrix} 0 & -\dot{\vartheta} & 0 \end{bmatrix}. \tag{2.88}$$

The rigid body is thus carrying out a planar rotation around the 2-axis with a negative direction of rotation, see also (2.53).

End of Example 2.3.

If we also apply (2.85) to (2.43), we then obtain after lengthy calculation the 3×1 rotation velocity vector as a function of the rotation parameters

$$\boldsymbol{\omega}(t) = \boldsymbol{d}\dot{\varphi} + \dot{\boldsymbol{d}} \sin\varphi + \tilde{\boldsymbol{d}} \cdot \dot{\boldsymbol{d}}(1 - \cos\varphi). \tag{2.89}$$

It is obvious that the rotation velocity depends not only on the temporal change $\dot{\varphi}(t)$ of the rotation angle but also on the temporal change $\dot{\boldsymbol{d}}(t)$ of the direction of the rotation axis. This underlines it clearly that finite rotation and instantaneous rotation have different properties.

If we insert half the rotation angle into (2.89), with the quaternions (2.44) we obtain the simplified relation

$$\boldsymbol{\omega}(t) = 2(q_0\dot{\boldsymbol{q}} - \dot{q}_0\boldsymbol{q} + \tilde{\boldsymbol{q}} \cdot \dot{\boldsymbol{q}}). \tag{2.90}$$

If we supplement (2.90) with the time derivative of (2.46)

$$q_0\dot{q}_0 + \boldsymbol{q} \cdot \dot{\boldsymbol{q}} = 0, \tag{2.91}$$

both equations can be merged into one 4×1 vector differential equation

$$\begin{bmatrix} 0 \\ \hline \boldsymbol{\omega}(t) \end{bmatrix} = 2\boldsymbol{Q}(q_0, \boldsymbol{q}) \cdot \begin{bmatrix} \dot{q}_0 \\ \hline \dot{\boldsymbol{q}} \end{bmatrix} = 2 \left[\begin{array}{c|c} q_0 & \boldsymbol{q} \\ \hline -\boldsymbol{q} & q_0\boldsymbol{E} + \tilde{\boldsymbol{q}} \end{array}\right] \cdot \begin{bmatrix} \dot{q}_0 \\ \hline \dot{\boldsymbol{q}} \end{bmatrix}, \tag{2.92}$$

which relates the rotation velocity with the quaternions. It should be noted in particular that the 4×4 coefficient matrix $\boldsymbol{Q}$ is orthogonal and thus nonsingular, so the inversion problem is easy to solve, see Table 2.2.

A further, intuitive way to calculate the rotation velocity vector is by using elementary rotations. For every elementary rotation there is one elementary rotational velocity. Following (2.57), we obtain

$$\boldsymbol{\omega}_{\alpha 1}(t) = \dot{\boldsymbol{\alpha}}_1(t) = [\dot{\alpha} \quad 0 \quad 0], \tag{2.93}$$

$$\boldsymbol{\omega}_{\beta 2}(t) = \dot{\boldsymbol{\beta}}_2(t) = [0 \quad \dot{\beta} \quad 0], \tag{2.94}$$

$$\boldsymbol{\omega}_{\gamma 3}(t) = \dot{\boldsymbol{\gamma}}_3(t) = [0 \quad 0 \quad \dot{\gamma}]. \tag{2.95}$$

Table 2.2 Kinematic differential equations

Sought coordinates	Rotation velocity in the spatially-fixed frame
9 direction cosines $\boldsymbol{S}$	$\dot{\boldsymbol{S}} = \tilde{\boldsymbol{\omega}} \cdot \boldsymbol{S}$
4 quaternions $[q_0\ \boldsymbol{q}]$	$\begin{bmatrix} \dot{q}_0 \\ \dot{\boldsymbol{q}} \end{bmatrix} = \frac{1}{2}\boldsymbol{Q}^T(q_0,\boldsymbol{q}) \cdot [0 \mid \boldsymbol{\omega}]$ $\begin{bmatrix} \dot{q}_0 \\ \dot{\boldsymbol{q}} \end{bmatrix} = \frac{1}{2}\left[\begin{array}{c\vert c} 0 & -\boldsymbol{\omega} \\ \hline \boldsymbol{\omega} & \tilde{\boldsymbol{\omega}} \end{array}\right] \cdot \begin{bmatrix} q_0 \\ \boldsymbol{q} \end{bmatrix}$
3 Cardano angle $\alpha(t), \beta(t), \gamma(t)$	$\dot{\boldsymbol{x}} = \boldsymbol{H}_R^{-1}(\boldsymbol{x}) \cdot \boldsymbol{\omega}$ $\boldsymbol{H}_R^{-1} = \begin{bmatrix} 1 & \sin\alpha\tan\beta & -\cos\alpha\tan\beta \\ 0 & \cos\alpha & \sin\alpha \\ 0 & -\dfrac{\sin\alpha}{\cos\beta} & \dfrac{\cos\alpha}{\cos\beta} \end{bmatrix}$
	Rotation velocity in the body-fixed frame 1
9 direction cosines $\boldsymbol{S}$	$\dot{\boldsymbol{S}} = \boldsymbol{S} \cdot {}_1\tilde{\boldsymbol{\omega}}$
4 quaternions $[q_0\ \boldsymbol{q}]$	$\begin{bmatrix} \dot{q}_0 \\ \dot{\boldsymbol{q}} \end{bmatrix} = \frac{1}{2}{}_1\boldsymbol{Q}^T(q_0,\boldsymbol{q}) \cdot [0 \mid {}_1\boldsymbol{\omega}]$ $\begin{bmatrix} \dot{q}_0 \\ \dot{\boldsymbol{q}} \end{bmatrix} = \frac{1}{2}\left[\begin{array}{c\vert c} 0 & -{}_1\boldsymbol{\omega} \\ \hline {}_1\boldsymbol{\omega} & -{}_1\tilde{\boldsymbol{\omega}} \end{array}\right] \cdot \begin{bmatrix} q_0 \\ \boldsymbol{q} \end{bmatrix}$
3 Cardano angle $\alpha(t), \beta(t), \gamma(t)$	$\dot{\boldsymbol{x}} = {}_1\boldsymbol{H}_R^{-1}(\boldsymbol{x}) \cdot {}_1\boldsymbol{\omega}$ ${}_1\boldsymbol{H}_R^{-1} = \begin{bmatrix} \dfrac{\cos\gamma}{\cos\beta} & -\dfrac{\sin\gamma}{\cos\beta} & 0 \\ \sin\gamma & \cos\gamma & 0 \\ -\cos\gamma\tan\beta & \sin\gamma\tan\beta & 1 \end{bmatrix}$

These elementary rotation velocities can be added vectorially, whereby the sequence of rotations and the transformations of the coordinate axes through previous rotations must be accounted for. This yields for the Euler angle

$$\boldsymbol{\omega}(t) = \dot{\boldsymbol{\psi}}_3(t) + \boldsymbol{\psi}_3(t) \cdot \dot{\boldsymbol{\vartheta}}_1(t) + \boldsymbol{\psi}_3(t) \cdot \boldsymbol{\vartheta}_1(t) \cdot \dot{\boldsymbol{\phi}}_3(t) \tag{2.96}$$

and for the Cardano angle

$$\boldsymbol{\omega}(t) = \dot{\boldsymbol{\alpha}}_1(t) + \boldsymbol{\alpha}_1(t) \cdot \dot{\boldsymbol{\beta}}_2(t) + \boldsymbol{\alpha}_1(t) \cdot \boldsymbol{\beta}_2(t) \cdot \dot{\boldsymbol{\gamma}}_3(t). \tag{2.97}$$

We can also represent rotation velocity vectors in a body-fixed frame, e.g. with the Cardano angles as

$$\boldsymbol{\omega}(t) = \boldsymbol{\gamma}_3^T(t) \cdot \boldsymbol{\beta}_2^T(t) \cdot \dot{\boldsymbol{\alpha}}_1(t) + \boldsymbol{\gamma}_3^T(t) \cdot \dot{\boldsymbol{\beta}}_2(t) + \dot{\boldsymbol{\gamma}}_3(t). \tag{2.98}$$

Evaluation of (2.97) yields, with the Cardano angles as generalized coordinates,

$$\boldsymbol{x}(t) = \begin{bmatrix} \alpha \; \beta \; \gamma \end{bmatrix}, \tag{2.99}$$

the relation

$$\boldsymbol{\omega}(t) = \boldsymbol{H}_R(\boldsymbol{x}) \cdot \dot{\boldsymbol{x}}(t) = \begin{bmatrix} 1 & 0 & \sin\beta \\ 0 & \cos\alpha & -\sin\alpha\cos\beta \\ 0 & \sin\alpha & \cos\alpha\cos\beta \end{bmatrix} \cdot \begin{bmatrix} \dot{\alpha} \\ \dot{\beta} \\ \dot{\gamma} \end{bmatrix}, \tag{2.100}$$

where the 3×3 Jacobian matrix $\boldsymbol{H}_R(\boldsymbol{x})$ of rotation has been introduced.

Calculation of the position or configuration from the rotation velocity is very important in dynamics. This can be done by integrating the corresponding differential equations. The rotation velocity is given in a body-fixed or spatially fixed frame, see Table 2.2. The kinematic differential equations of the direction cosines and quaternions are overdetermined. The constraints shown in Table 2.1 are included in differentiated form in the differential equations, although a first integral is known, namely the constraints themselves. This can lead to numerical difficulties, i.e. the constraints can be violated in case of longer integrations. It is therefore wise to provide a correction method that performs a normalization in accordance with (2.32) or (2.46) after each integration step. This normalization is undertaken automatically for the differential equation of the Cardano angle as well as for all other elementary rotations. However, then the functional matrix $\boldsymbol{H}_R$ is no longer regular in the singular configurations. Yet these singularities can be avoided by using complementary rotation angles, an adjustment which requires a higher level of programming effort.

Example 2.4 (Integration of the Direction Cosines). Let the rotation velocity vector (2.88) of the rotation be around the negative 2-axis and the initial condition be $\boldsymbol{S}(t = t_0) = \boldsymbol{S}_0$. According to Table 2.2, the differential equations of the direction cosines are then written $\dot{\boldsymbol{S}} = \tilde{\boldsymbol{\omega}} \cdot \boldsymbol{S}$ or

$$\begin{array}{lll} \dot{S}_{11} = \omega_2 S_{31}, & \dot{S}_{12} = \omega_2 S_{32}, & \dot{S}_{13} = \omega_2 S_{33}, \\ \dot{S}_{21} = 0, & \dot{S}_{22} = 0, & \dot{S}_{23} = 0, \\ \dot{S}_{31} = -\omega_2 S_{11}, & \dot{S}_{32} = -\omega_2 S_{12}, & \dot{S}_{33} = -\omega_2 S_{13}. \end{array} \tag{2.101}$$

If we now consider that linear, time-variant differential equation systems of the form

$$\begin{bmatrix} \dot{x}_1(t) \\ \dot{x}_2(t) \end{bmatrix} = \begin{bmatrix} 0 & \omega(t) \\ -\omega(t) & 0 \end{bmatrix} \cdot \begin{bmatrix} x_1(t) \\ x_2(t) \end{bmatrix}, \tag{2.102}$$

$$\begin{bmatrix} x_1(t = t_0) \\ x_2(t = t_0) \end{bmatrix} = \begin{bmatrix} x_{10} \\ x_{20} \end{bmatrix} \tag{2.103}$$

have the general solution

$$\begin{bmatrix} x_1(t) \\ x_2(t) \end{bmatrix} = \begin{bmatrix} \cos\left(\int_{t_0}^{t} \omega dt\right) & \sin\left(\int_{t_0}^{t} \omega dt\right) \\ -\sin\left(\int_{t_0}^{t} \omega dt\right) & \cos\left(\int_{t_0}^{t} \omega dt\right) \end{bmatrix} \cdot \begin{bmatrix} x_{10} \\ x_{20} \end{bmatrix}, \tag{2.104}$$

then we obtain from (2.101) with $\omega_2 = -\dot{\vartheta}$,

$$\boldsymbol{S}(t) = \begin{bmatrix} \cos\vartheta & 0 & -\sin\vartheta \\ 0 & 1 & 0 \\ \sin\vartheta & 0 & \cos\vartheta \end{bmatrix} \cdot \begin{bmatrix} S_{110} & S_{120} & S_{130} \\ S_{210} & S_{220} & S_{230} \\ S_{310} & S_{320} & S_{330} \end{bmatrix}. \tag{2.105}$$

If $\boldsymbol{S}_0 = \boldsymbol{E}$, we again obtain the rotation tensor (2.49).

If the differential equations (2.101) are solved not analytically but numerically, integration errors can destroy the orthogonality. For example, if we allow one integration error ε in a solution of the differential equation (2.101),

$$S_{11} = \cos\vartheta + \varepsilon, \tag{2.106}$$

then the corresponding condition of orthogonality is no longer satisfied. For $\boldsymbol{S}_0 = \boldsymbol{E}$ we obtain for example

$$S_{11}^2 + S_{21}^2 + S_{31}^2 = 1 + 2\varepsilon\cos\vartheta \neq 1. \tag{2.107}$$

A non-orthogonal rotation tensor corresponds however to the deformation gradient of a nonrigid body. For this reason, orthogonality must constantly be checked and enforced.

End of Example 2.4.

Example 2.5 (Integration of the Cardano Angle). The rotation shown in Example 2.4 will now be described with Cardano angles. In Table 2.2 we find

$$\dot{\alpha} = \omega_2 \sin\alpha \tan\beta, \qquad \dot{\beta} = \omega_2 \cos\alpha, \qquad \dot{\gamma} = -\omega_2 \frac{\sin\alpha}{\cos\beta}. \tag{2.108}$$

This nonlinear system of differential equations can now be solved in a closed form with the initial condition $\alpha_0 = 0$, $\beta_0 = 0$, $\gamma_0 = 0$

$$\alpha(t) = 0, \qquad \beta(t) = \int \omega_2 dt = -\vartheta, \qquad \gamma(t) = 0. \tag{2.109}$$

The rotation tensor (2.49) has again been determined, and orthogonality has by definition been preserved.

On the other hand, there is a singularity for $\alpha_0 = -\gamma_0 = 0$ and $\beta_0 = \pi/2$. This can be rectified by using complementary Cardano angles (2.64). According to (2.71), (2.72), and (2.74), the corresponding initial conditions are $\overline{\alpha}_0 = 0$,

$\overline{\beta}_0 = \pi/2$, and $\overline{\gamma}_0 = 0$ and the nonlinear time-variant system of differential equations has the form

$$\dot{\overline{\alpha}} = -\omega_2 \sin\overline{\alpha}\cot\overline{\beta}, \qquad \dot{\overline{\beta}} = \omega_2 \cos\overline{\alpha}, \qquad \dot{\overline{\gamma}} = \omega_2 \frac{\sin\overline{\alpha}}{\sin\overline{\beta}} \tag{2.110}$$

with the solution

$$\overline{\alpha}(t) = 0, \qquad \overline{\beta}(t) = \frac{\pi}{2} + \int \omega_2 dt, \qquad \overline{\gamma}(t) = 0. \tag{2.111}$$

We can see that the transition from Cardano angles to complementary Cardano angles was even possible in one singular position. But this should be avoided for numerical reasons.

End of Example 2.5.

This concludes our discussion of rotational velocity. According to (2.86), the velocity of a rigid body is given completely by the translational velocity $\boldsymbol{v}(t)$ of a mass point $P(t)$ and by the rotational velocity $\boldsymbol{\omega}(t)$ of the body, which is equal at every point. These velocities can however also be expressed in accordance with (2.6) and (2.100) by the generalized coordinates of the 6×1 position vector (2.81) as

$$\begin{aligned} \boldsymbol{v}(t) &= \boldsymbol{v}(\boldsymbol{x},\dot{\boldsymbol{x}}) = [\boldsymbol{H}_T(\boldsymbol{x}) \;\; \boldsymbol{0}] \cdot \dot{\boldsymbol{x}}(t) = \overline{\boldsymbol{H}}_T(\boldsymbol{x}) \cdot \dot{\boldsymbol{x}}(t), \\ \boldsymbol{\omega}(t) &= \boldsymbol{\omega}(\boldsymbol{x},\dot{\boldsymbol{x}}) = [\boldsymbol{0} \;\; \boldsymbol{H}_R(\boldsymbol{x})] \cdot \dot{\boldsymbol{x}}(t) = \overline{\boldsymbol{H}}_R(\boldsymbol{x}) \cdot \dot{\boldsymbol{x}}(t), \end{aligned} \tag{2.112}$$

whereby the 3×6 functional matrices can be obtained from (2.6) and (2.100) by extending the matrices by zero submatrices. Formally, it yields according to (2.7)

$$\overline{\boldsymbol{H}}_T(\boldsymbol{x}) = \frac{\partial \boldsymbol{r}(\boldsymbol{x})}{\partial \boldsymbol{x}}, \qquad \overline{\boldsymbol{H}}_R(\boldsymbol{x}) = \frac{\partial \boldsymbol{s}(\boldsymbol{x})}{\partial \boldsymbol{x}}. \tag{2.113}$$

It should be noted that, in the second formula of (2.113), the infinitesimal instantaneous rotation from (2.85) must be used. The transition from (2.82) to (2.113) is therefore somewhat laborious and must be undertaken using the skew-symmetric tensor of the infinitesimal instantaneous rotation tensor

$$\frac{\partial \tilde{s}_{\alpha\beta}}{\partial x_\delta} = \frac{\partial S_{\alpha\gamma}}{\partial x_\delta} S_{\beta\gamma}, \qquad \alpha,\beta,\gamma = 1(1)3, \quad \delta = 1(1)6. \tag{2.114}$$

Equation (2.114) is best analyzed with a formula manipulation program. However, the Jacobian matrix $\overline{\boldsymbol{H}}_R(\boldsymbol{x})$ in (2.113) can also be obtained descriptively with the help of elementary rotations from (2.100).

The current acceleration of the rigid body is given by a further time derivative from (2.83)

$$\boldsymbol{a}(\boldsymbol{\rho},t) = \frac{d}{dt}\boldsymbol{v}(\boldsymbol{\rho},t) = \ddot{\boldsymbol{r}}_1(t) + \ddot{\boldsymbol{S}}(t) \cdot \boldsymbol{\rho}. \tag{2.115}$$

If we again consider (2.79), we obtain

$$\boldsymbol{a}(\boldsymbol{\rho},t) = \ddot{\boldsymbol{r}}_1(t) + \ddot{\boldsymbol{S}}(t)\cdot\boldsymbol{S}^T(t)\cdot\boldsymbol{r}_P(\boldsymbol{\rho},t). \tag{2.116}$$

We again recognize the translational acceleration (2.12) as the first term, while the second term denotes the rotational acceleration, yielding

$$\begin{aligned}\ddot{\boldsymbol{S}}\cdot\boldsymbol{S}^T &= \ddot{\boldsymbol{S}}\cdot\boldsymbol{S}^T + \dot{\boldsymbol{S}}\cdot\dot{\boldsymbol{S}}^T - \dot{\boldsymbol{S}}\cdot\dot{\boldsymbol{S}}^T = \ddot{\boldsymbol{S}}\cdot\boldsymbol{S}^T + \dot{\boldsymbol{S}}\cdot\dot{\boldsymbol{S}}^T - \dot{\boldsymbol{S}}\cdot\boldsymbol{S}^T\cdot\boldsymbol{S}\cdot\dot{\boldsymbol{S}}^T\\ &= \dot{\tilde{\boldsymbol{\omega}}}(t) + \tilde{\boldsymbol{\omega}}(t)\cdot\tilde{\boldsymbol{\omega}}(t),\end{aligned} \tag{2.117}$$

where the definition (2.85) of angular velocity and its derivatives $\dot{\tilde{\boldsymbol{\omega}}} = \ddot{\boldsymbol{S}}\cdot\boldsymbol{S}^T + \dot{\boldsymbol{S}}\cdot\dot{\boldsymbol{S}}^T$ are applied. If we now introduce the 3×1 rotational acceleration vector

$$\boldsymbol{\alpha}(t) = \dot{\boldsymbol{\omega}}(t), \tag{2.118}$$

we then obtain

$$\boldsymbol{a}(\boldsymbol{\rho},t) = \ddot{\boldsymbol{r}}_1(t) + \left[\tilde{\boldsymbol{\alpha}}(t) + \tilde{\boldsymbol{\omega}}(t)\cdot\tilde{\boldsymbol{\omega}}(t)\right]\cdot\boldsymbol{r}_p(\boldsymbol{\rho},t). \tag{2.119}$$

The acceleration of a rigid body is thus given by the translational acceleration $\boldsymbol{a}_1(t)$ of the mass point P_1, its rotational acceleration $\boldsymbol{\alpha}(t)$ and the square of its rotational velocity $\boldsymbol{\omega}(t)$.

Just like in (2.13), the accelerations can be expressed with generalized coordinates

$$\boldsymbol{a}(t) = \boldsymbol{a}(\boldsymbol{x},\dot{\boldsymbol{x}},\ddot{\boldsymbol{x}}) = \overline{\boldsymbol{H}}_T(\boldsymbol{x})\cdot\ddot{\boldsymbol{x}}(t) + \left(\frac{\partial\overline{\boldsymbol{H}}_T(\boldsymbol{x})}{\partial\boldsymbol{x}}\cdot\dot{\boldsymbol{x}}(t)\right)\cdot\dot{\boldsymbol{x}}(t), \tag{2.120}$$

$$\boldsymbol{\alpha}(t) = \boldsymbol{\alpha}(\boldsymbol{x},\ddot{\boldsymbol{x}},\dot{\boldsymbol{x}}) = \overline{\boldsymbol{H}}_R(\boldsymbol{x})\cdot\ddot{\boldsymbol{x}}(t) + \left(\frac{\partial\overline{\boldsymbol{H}}_R(\boldsymbol{x})}{\partial\boldsymbol{x}}\cdot\dot{\boldsymbol{x}}(t)\right)\cdot\dot{\boldsymbol{x}}(t). \tag{2.121}$$

We thus obtain, in accordance with (2.112), the 3×6 functional matrices $\overline{\boldsymbol{H}}_T$ and $\overline{\boldsymbol{H}}_R$ and their derivatives.

A free system of p rigid bodies has $6p$ degrees of freedom, which are described by the $6p\times 1$ position vector $\boldsymbol{x}(t)$ of the generalized coordinates of the overall system. Applying (2.82), for the ith body

$$\boldsymbol{r}_i(t) = \boldsymbol{r}_i(\boldsymbol{x}), \qquad \boldsymbol{S}_i(t) = \boldsymbol{S}_i(\boldsymbol{x}) \tag{2.122}$$

is true. We can likewise apply (2.112)–(2.114) and (2.121) to the ith body.

2.1.3 Kinematics of a Continuum

Like the rigid body, the continuum is a model of mechanics. The distances between the mass points of a continuum are not constant like those of a rigid body, however. A continuum in the deformation process is thus subject not only to a translation and a rotation but also to strain. However, strain is generally small in elastic materials, so we can usually work with linear relations. Fluids, which are subject to major strains, or plastic materials will not be dealt with in this book. Strains arising in a continuum also permit the calculation of internal forces and stresses, which are of crucial importance in strength tests. Nevertheless, the use of continuum models in dynamics is not always mandatory. Frequently, motions are calculated with a rigid body model and the strength tests – under consideration of inertia forces – are carried out using static methods. In order to describe free continua kinematically, it is again sufficient to consider a single body as was the case in the previous sections.

In order to describe the configuration of a continuum K mathematically, Fig. 2.4 and Eqs. (2.25)–(2.30) can be adopted without any alterations. The deformation gradient $\boldsymbol{F}(\boldsymbol{\rho},t)$ is now no longer orthogonal, however. But it can, like any second-order tensor, undergo a polar decomposition

$$\boldsymbol{F}(\boldsymbol{\rho},t) = \overline{\boldsymbol{S}}(\boldsymbol{\rho},t) \cdot \boldsymbol{U}(\boldsymbol{\rho},t), \tag{2.123}$$

where we see not only the location-dependent, proper orthogonal 3×3 rotation tensor

$$\overline{\boldsymbol{S}}^T(\boldsymbol{\rho},t) = \overline{\boldsymbol{S}}^{-1}(\boldsymbol{\rho},t) \tag{2.124}$$

but also the equally location-dependent, symmetric and positive definite 3×3 right stretch tensor

$$\boldsymbol{U}^T(\boldsymbol{\rho},t) = \boldsymbol{U}(\boldsymbol{\rho},t), \tag{2.125}$$

which is a measure of strain. Proof of the properties mentioned can be found e.g. in Lai, Rubin, Krempl [31] or Becker and Bürger [9] and will not be repeated here. From the 3×3 right stretch tensor we obtain the Green strain tensor

$$\boldsymbol{G} = \frac{1}{2}(\boldsymbol{U} \cdot \boldsymbol{U} - \boldsymbol{E}) = \frac{1}{2}(\boldsymbol{F}^T \cdot \boldsymbol{F} - \boldsymbol{E}), \tag{2.126}$$

which is also symmetric. With (2.123) and (2.125) we can also write the deformation gradient as

$$\boldsymbol{F} = \overline{\boldsymbol{S}} \cdot (\boldsymbol{E} + 2\boldsymbol{G})^{\frac{1}{2}}. \tag{2.127}$$

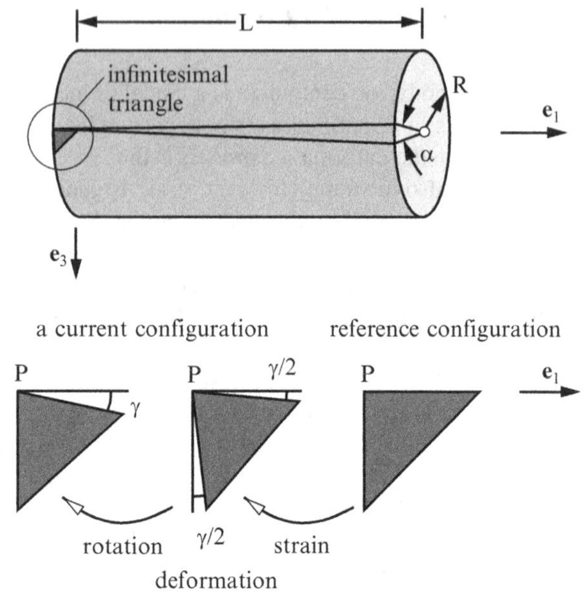

Fig. 2.9 Deformation of a round bar

Further information on the calculation of the roots of a matrix can be found in Zurmühl and Falk [69]. In the case of a rigid body, the Green strain tensor disappears, and from $\boldsymbol{U} = \boldsymbol{E}$ follows $\boldsymbol{G} = \boldsymbol{0}$, as a result of which (2.127) reverts to (2.33).

Example 2.6 (Strain of a Twisted Round Bar). The current configuration of a twisted round bar, Fig. 2.9, is described by point P with the 3×1 location vector

$$\boldsymbol{r}(\boldsymbol{\rho},t) = \begin{bmatrix} \rho_1 \\ \rho_2 - \alpha\rho_3 \\ \rho_3 + \alpha\rho_2 \end{bmatrix}, \qquad \alpha(\rho_1,t) \ll 1, \tag{2.128}$$

where the 3×1 location vector $\boldsymbol{\rho}$ denotes the mass points in the reference configurations. The small angle α is a function of location and time, its location-dependence being restricted to the longitudinal direction of the bar.

According to (2.28), the deformation gradient is

$$\boldsymbol{F} = \begin{bmatrix} 1 & 0 & 0 \\ -\alpha'\rho_3 & 1 & -\alpha \\ \alpha'\rho_2 & \alpha & 1 \end{bmatrix}, \qquad \alpha' = \frac{\partial\alpha}{\partial\rho_1} \ll 1 \tag{2.129}$$

and the square of the right stretch tensor without taking quadratically small magnitudes into account, results in

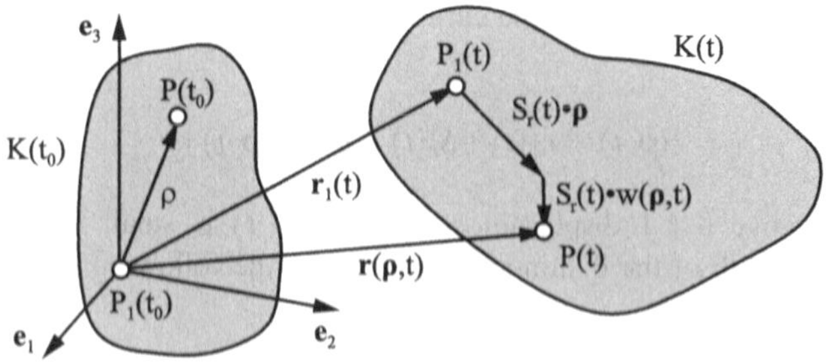

Fig. 2.10 Free motion of a linear-elastic body

$$\boldsymbol{U}\cdot\boldsymbol{U}=\boldsymbol{F}^T\cdot\boldsymbol{F}=\begin{bmatrix}1 & -\alpha'\rho_3 & \alpha'\rho_2\\ -\alpha'\rho_3 & 1 & 0\\ \alpha'\rho_2 & 0 & 1\end{bmatrix}. \tag{2.130}$$

If $\alpha \ll 1$, then $\boldsymbol{U}$ from (2.130) can be written in a closed form

$$\boldsymbol{U}=\begin{bmatrix}1 & -\frac{1}{2}\alpha'\rho_3 & \frac{1}{2}\alpha'\rho_2\\ -\frac{1}{2}\alpha'\rho_3 & 1 & 0\\ \frac{1}{2}\alpha'\rho_2 & 0 & 1\end{bmatrix}. \tag{2.131}$$

With (2.123) we now find the rotation tensor

$$\overline{\boldsymbol{S}}=\begin{bmatrix}1 & \frac{1}{2}\alpha'\rho_3 & -\frac{1}{2}\alpha'\rho_2\\ -\frac{1}{2}\alpha'\rho_3 & 1 & -\alpha\\ \frac{1}{2}\alpha'\rho_2 & \alpha & 1\end{bmatrix} \tag{2.132}$$

and with (2.126) we obtain for the linearized Green strain tensor

$$\boldsymbol{G}_{lin}=\frac{\alpha'}{2}\begin{bmatrix}0 & -\rho_3 & \rho_2\\ -\rho_3 & 0 & 0\\ \rho_2 & 0 & 0\end{bmatrix}. \tag{2.133}$$

The relation $R\alpha(\rho_1)=\gamma\rho_1$ applies for cases of static strain, see Fig. 2.9. If we consider the deformation of an infinitesimal triangle at the mass point $\boldsymbol{\rho}=[0\ R\ 0]$, we determine from Fig. 2.9 that in addition to strain, characterized by a pure change of angle, there is also a rotation of the infinitesimal triangle. This confirms the assertion of (2.127) concerning simultaneously possible rotations and strains in nonrigid bodies.

End of Example 2.6.

If we now note that elastic strain in elastic materials are usually small in relation to rigid body motion, then the above relations can generally be linearized, see Fig. 2.10.

Now the following is true for the current configuration with rigid body rotation $\boldsymbol{S}_r(t)$,

$$\boldsymbol{r}(\boldsymbol{\rho},t) = \boldsymbol{r}_1(t) + \boldsymbol{S}_r(t)\cdot\left[\boldsymbol{\rho} + \boldsymbol{w}(\boldsymbol{\rho},t)\right], \tag{2.134}$$

where the relative 3×1 displacement vector $\boldsymbol{w}(\boldsymbol{\rho},t)$ is small in relation to a characteristic length of the continuum. In addition, the following relation is valid in (2.134)

$$\boldsymbol{w}(\boldsymbol{0},t) = \boldsymbol{0}, \tag{2.135}$$

which specifies the location vector $\boldsymbol{r}_1(t)$ to the reference point P_1. With the associated 3×3 displacement gradient

$$\boldsymbol{F}_w(\boldsymbol{\rho},t) = \frac{\partial \boldsymbol{w}}{\partial \boldsymbol{\rho}} \tag{2.136}$$

the linearized deformation gradient is

$$\boldsymbol{F}_{lin} = \overline{\boldsymbol{S}}\cdot(\boldsymbol{E} + \boldsymbol{G}_{lin}), \tag{2.137}$$

where the relations

$$\boldsymbol{G}_{lin} = \frac{1}{2}(\boldsymbol{F}_w + \boldsymbol{F}_w^T), \qquad \overline{\boldsymbol{S}} = \boldsymbol{S}_r(t)\cdot\boldsymbol{S}_w(\boldsymbol{\rho},t), \tag{2.138}$$

$$\boldsymbol{S}_w = \boldsymbol{E} + \frac{1}{2}(\boldsymbol{F}_w - \boldsymbol{F}_w^T), \qquad \boldsymbol{S}_w(\boldsymbol{0},t) = \boldsymbol{E}, \tag{2.139}$$

must be taken into account. In the linear case, we thus obtain the linear 3×3 Green strain tensor $\boldsymbol{G}_{lin}$ and the 3×3 tensor $\boldsymbol{S}_w$ of relative rotation by simple decomposition of the displacement gradient $\boldsymbol{F}_w$ into its symmetric and skew-symmetric components, see (2.138) and (2.139). It should also be mentioned that the tensor $\boldsymbol{S}_w(\boldsymbol{\rho},t)$ of relative motion, as opposed to the rotation tensor $\boldsymbol{S}_r(t)$, is dependent on location and time. According to (2.138), the total rotation $\overline{\boldsymbol{S}}(\boldsymbol{\rho},t)$ is composed of the rigid body rotation $\boldsymbol{S}_r(t)$ and the relative rotation $\boldsymbol{S}_w(\boldsymbol{\rho},t)$.

The linearized Green strain tensor has, for reasons of symmetry, only six essential elements

$$\boldsymbol{G}_{lin} = \begin{bmatrix} \varepsilon_{11} & \varepsilon_{12} & \varepsilon_{31} \\ \varepsilon_{12} & \varepsilon_{22} & \varepsilon_{23} \\ \varepsilon_{31} & \varepsilon_{23} & \varepsilon_{33} \end{bmatrix}, \tag{2.140}$$

which can also be merged into a 6×1 strain vector

$$\boldsymbol{e} = \begin{bmatrix} \varepsilon_{11} & \varepsilon_{22} & \varepsilon_{33} & \gamma_{12} & \gamma_{23} & \gamma_{31} \end{bmatrix}. \tag{2.141}$$

We call $\varepsilon_{\alpha\alpha}$, $\alpha = 1(1)3$ normal strains or elongations. The secondary diagonal elements $\varepsilon_{12}, \varepsilon_{23}, \varepsilon_{31}$ are called shear strains. Here, in the strain vector $\gamma_{12} = 2\varepsilon_{12}$, $\gamma_{23} = 2\varepsilon_{23}$, $\gamma_{31} = 2\varepsilon_{31}$ appear, which are referred to as slidings and describe the changes in one of the 90° angles in the reference configuration. The strains and shear stresses are not mutually independent however, since they are calculated from the three coordinates of the displacement vector $\boldsymbol{w}$. Yet the conditions of compatibility are now no longer denoted by algebraic equations but rather by differential equations. If we now introduce the 6×3 differential operator matrix,

$$\mathscr{V} = \begin{bmatrix} \partial/\partial\rho_1 & 0 & 0 \\ 0 & \partial/\partial\rho_2 & 0 \\ 0 & 0 & \partial/\partial\rho_3 \\ \partial/\partial\rho_2 & \partial/\partial\rho_1 & 0 \\ 0 & \partial/\partial\rho_3 & \partial/\partial\rho_2 \\ \partial/\partial\rho_3 & 0 & \partial/\partial\rho_1 \end{bmatrix}, \tag{2.142}$$

we can also calculate the strain vector directly from the displacement vector,

$$\boldsymbol{e} = \mathscr{V} \cdot \boldsymbol{w}. \tag{2.143}$$

The algorithms of matrix multiplication are applicable for the differential operator matrix $\mathscr{V}$, as shown in Sect. A.3.

As a consequence of linearization, the rotation tensor (2.139) has only three essential elements,

$$\boldsymbol{S}_w = \begin{bmatrix} 1 & -\gamma & \beta \\ \gamma & 1 & -\alpha \\ -\beta & \alpha & 1 \end{bmatrix}, \tag{2.144}$$

which correspond to the small Cardano angles α, β, γ. The essential elements of (2.144) can be merged in the 3×1 rotation vector

$$\boldsymbol{s} = \begin{bmatrix} \alpha & \beta & \gamma \end{bmatrix} \tag{2.145}$$

and with the 3×3 differential operator matrix of elastic strain,

$$\mathscr{D} = \frac{1}{2} \begin{bmatrix} 0 & -\partial/\partial\rho_3 & \partial/\partial\rho_2 \\ \partial/\partial\rho_3 & 0 & -\partial/\partial\rho_1 \\ -\partial/\partial\rho_2 & \partial/\partial\rho_1 & 0 \end{bmatrix}, \tag{2.146}$$

determined from the displacement vector,

$$\boldsymbol{s} = \mathscr{D} \cdot \boldsymbol{w}. \tag{2.147}$$

The rotation vector (2.145) plays an important role in the mechanics of polar continua, among which we can include the Bernoulli beam, too. A polar continuum is composed of mass points, which can execute both displacements and rotations and is also known as a Cosserat continuum.

A nonrigid continuum has an infinite number of degrees of freedom since it comprises infinitely many free mass points. This is also reflected in the fact that the deformation gradient is dependent not only on time but also on the location or material coordinates, respectively, of the mass points. A solution method often used in linear continuum mechanics exploits this fact in conjunction with the principles of separation and superposition,

$$\boldsymbol{w}(\boldsymbol{\rho},t) = \boldsymbol{A}(\boldsymbol{\rho}) \cdot \boldsymbol{x}(t), \tag{2.148}$$

where the $3 \times f$ matrix $\boldsymbol{A}(\boldsymbol{\rho})$ of the relative shape functions and the $f \times 1$ position vector $\boldsymbol{x}(t)$ of the generalized coordinates appear with $f \to \infty$. The approach (2.148), which does not always lead to the desired aim, thus in particular indicates the infinitely many degrees of freedom of the continuum. With (2.143), we also obtain for the strain vector

$$\boldsymbol{e}(\boldsymbol{\rho},t) = \boldsymbol{B}(\boldsymbol{\rho}) \cdot \boldsymbol{x}(t) \tag{2.149}$$

with the $6 \times f$ matrix $\boldsymbol{B}(\boldsymbol{\rho})$ of the strain functions,

$$\boldsymbol{B}(\boldsymbol{\rho}) = \mathscr{V} \cdot \boldsymbol{A}(\boldsymbol{\rho}). \tag{2.150}$$

For small elements of a continuum, it is also sufficient to choose with a finite number of generalized coordinates, such as are utilized in the finite element method, see Chap. 6. Furthermore, if we assume linear kinematics in the rigid body motion with reference to point P_1, we obtain from (2.134) and (2.148) for the displacement vector

$$\boldsymbol{r}(\boldsymbol{\rho},t) = \boldsymbol{\rho} + \boldsymbol{C}(\boldsymbol{\rho}) \cdot \boldsymbol{x}(t), \tag{2.151}$$

yielding the $3 \times f$ matrix $\boldsymbol{C}(\boldsymbol{\rho})$ of the absolute shape functions and a corresponding $f \times 1$ position vector $\boldsymbol{x}(t)$. Using the finite element method, the $f \times 1$ position vector $\boldsymbol{x}(t)$ is determined by means of the Cartesian coordinates of single mass points $P_j, j = 1,2,3,\ldots$,

$$\boldsymbol{x}(t) = \left[\boldsymbol{r}(\boldsymbol{\rho}_1,t)\; \boldsymbol{r}(\boldsymbol{\rho}_2,t)\; \boldsymbol{r}(\boldsymbol{\rho}_3,t) \ldots\right]. \tag{2.152}$$

In the case of continuous systems on the other hand, often the generalized coordinates belonging to the eigenforms are merged in the position vector, see Chap. 7.

The current velocity of a point in the continuum is determined by the time derivative of (2.25)

$$\boldsymbol{v}(\boldsymbol{\rho},t) = \frac{d}{dt}\boldsymbol{r}(\boldsymbol{\rho},t). \tag{2.153}$$

We can obtain additional information if deformation is noted in accordance with (2.28), (2.29),

$$\boldsymbol{v}(\boldsymbol{\rho}+d\boldsymbol{\rho}) = \boldsymbol{v}(\boldsymbol{\rho}) + \dot{\boldsymbol{F}}(\boldsymbol{\rho})\cdot\boldsymbol{F}^{-1}(\boldsymbol{\rho})\cdot d\boldsymbol{r}(\boldsymbol{\rho}). \tag{2.154}$$

This yields the tensor of the spatial velocity gradient,

$$\boldsymbol{L} = \dot{\boldsymbol{F}}\cdot\boldsymbol{F}^{-1} = \frac{\partial\boldsymbol{v}(\boldsymbol{r})}{\partial\boldsymbol{r}}, \tag{2.155}$$

which can be decomposed into symmetric and skew-symmetric components

$$\boldsymbol{L} = \boldsymbol{D}+\boldsymbol{W}, \qquad \boldsymbol{D} = \frac{1}{2}(\boldsymbol{L}+\boldsymbol{L}^T), \qquad \boldsymbol{W} = \frac{1}{2}(\boldsymbol{L}-\boldsymbol{L}^T). \tag{2.156}$$

Here, $\boldsymbol{D}$ denotes the symmetric 3×3 strain velocity tensor, while $\boldsymbol{W}$ describes the skew-symmetric 3×3 rotation velocity tensor. Comparing (2.84) and (2.154), we can see clearly that, due to (2.33), the strain velocity tensor vanishes as expected in the case of a rigid body. In the linear case, we obtain from (2.137) to (2.139) with (2.155) and (2.156), ignoring quadratically small elements

$$\boldsymbol{D} = \overline{\boldsymbol{S}}\cdot\dot{\boldsymbol{G}}_w\cdot\overline{\boldsymbol{S}}^T, \qquad \boldsymbol{W} = \dot{\overline{\boldsymbol{S}}}\cdot\overline{\boldsymbol{S}}^T. \tag{2.157}$$

If we finally base our examination on the approach in (2.151), we then have

$$\boldsymbol{v}(\boldsymbol{\rho},t) = \boldsymbol{C}(\boldsymbol{\rho})\cdot\dot{\boldsymbol{x}}(t) \tag{2.158}$$

for the current velocity.

The current acceleration of a point in the continuum is taken from (2.153) via mass derivation of the velocity

$$\boldsymbol{a}(\boldsymbol{\rho},t) = \frac{d}{dt}\boldsymbol{v}(\boldsymbol{\rho},t) = \frac{\partial}{\partial t}\boldsymbol{v}(\boldsymbol{r},t) = \frac{\partial\boldsymbol{v}(\boldsymbol{r},t)}{\partial\boldsymbol{r}}\cdot\boldsymbol{v} + \frac{\partial\boldsymbol{v}(\boldsymbol{r},t)}{\partial t}. \tag{2.159}$$

In this case, first the inverse function (2.27) was utilized, followed by a separation of acceleration into a convective component (spatial velocity gradient) and a local component. Furthermore, from (2.158) we obtain for the linear kinematics

$$\boldsymbol{a}(\boldsymbol{\rho},t) = \boldsymbol{C}(\boldsymbol{\rho})\cdot\ddot{\boldsymbol{x}}(t). \tag{2.160}$$

The kinematics of the free continuum is now also concluded.

Example 2.7 (Velocity and Acceleration of a Round Bar). From the current configuration (2.128), we obtain the vectors from the time derivative, which will in the following be indicated by a point $(\dot{\cdot})$,

$$\boldsymbol{v}(\boldsymbol{\rho},t) = \begin{bmatrix} 0 \\ \dot{\alpha}\rho_3 \\ -\dot{\alpha}\rho_2 \end{bmatrix}, \qquad \dot{\alpha} = \frac{d\alpha(\rho_1,t)}{dt}, \tag{2.161}$$

$$\boldsymbol{a}(\boldsymbol{\rho},t) = \begin{bmatrix} 0 \\ \ddot{\alpha}\rho_3 \\ -\ddot{\alpha}\rho_2 \end{bmatrix}, \qquad \ddot{\alpha} = \frac{d^2\alpha(\rho_1,t)}{dt^2}. \tag{2.162}$$

If we now observe the inverse function of (2.128) or the material coordinates, respectively,

$$\boldsymbol{\rho}(\boldsymbol{r},t) = \begin{bmatrix} r_1 \\ r_2 - \alpha r_3 \\ r_3 + \alpha r_2 \end{bmatrix}, \tag{2.163}$$

we then see that (2.161) and (2.162) are also valid in spatial coordinates. Also, it can be shown quite generally that there are no differences between representations with material and spatial coordinates in linear kinematics.

End of Example 2.7.

2.2 Holonomic Systems

Constrained systems differ from free systems in that the freedom of motion of one or more position variables is limited by mechanical constraints. In engineering, holonomic constraints are realized by means of ideal, i.e. inflexible guides, joints, levers, bearings, rods, and other connections. The constraints between particular machine elements permit the engineer to arrive at a certain total motion in order to solve an engineering problem. On the other hand, constraints also serve to break down a complicated total motion into simple sub-motions, which can then be controlled independently of each other. For an industrial robot, see Fig. 2.11, each degree of freedom is normally assigned one rigid body and a drive motor.

When defining holonomic systems, it is sensible to allow free systems as a special case. This opens up more possibilities for the mathematical description of free systems. The representation of a free system in the form of a holonomic system means nothing more than an additional coordinate transformation. The number of degrees of freedom is unaffected by this, as is the mechanical issue of lacking constraints.

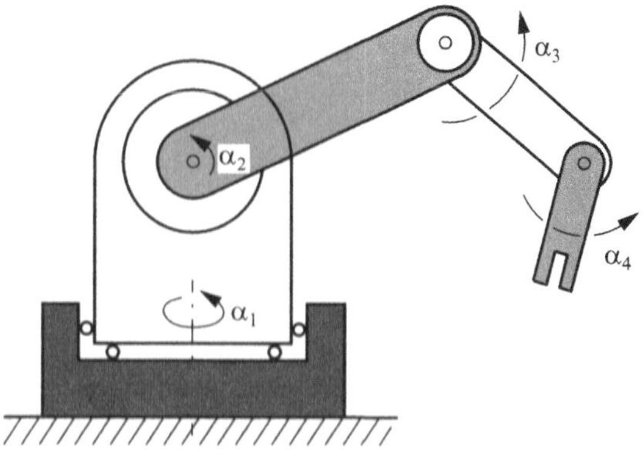

Fig. 2.11 Industrial articulated robot with 4 degrees of freedom

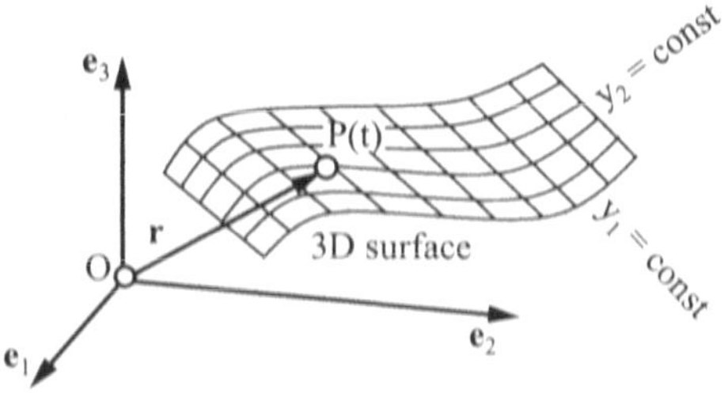

Fig. 2.12 Motion of a point on a three-dimensional surface

2.2.1 *Mass Point Systems*

Constraints will first be discussed again using the example of a single point. The motion of a mass point $P(t)$ can be restricted due to being confined to a surface or a path. Translational displacement on a surface that is changeable in time, see Fig. 2.12, can be represented uniquely by means of two generalized coordinates $y_1(t), y_2(t)$ corresponding to the two degrees of freedom,

$$\boldsymbol{r}(t) = \boldsymbol{r}(\boldsymbol{x}) = \boldsymbol{r}(y_1, y_2, t). \tag{2.164}$$

A surface in space is described by a scalar, algebraic and usually nonlinear equation,

$$\phi(\boldsymbol{x}, t) = 0, \tag{2.165}$$

where $\boldsymbol{x}(t)$ is the 3×1 position vector of the free point. This provides us with an implicit constraint onto a surface. With (2.164), we can also represent the constraint explicitly,

$$\boldsymbol{x} = \boldsymbol{x}(y_1, y_2, t). \tag{2.166}$$

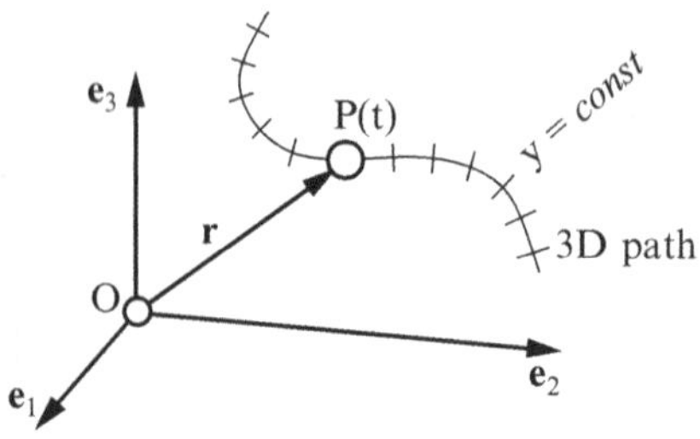

Fig. 2.13 Motion of a point along a three-dimensional path

Both representations are equivalent. By means of (2.166), the reduction of the order of the position vector due to the constraint is obvious. Translation along a time-varying path, see Fig. 2.13, has only one degree of freedom with one generalized coordinate $y(t)$. The following applies

$$\boldsymbol{r}(t) = \boldsymbol{r}(\boldsymbol{x}) = \boldsymbol{r}(y,t). \tag{2.167}$$

A path in space is given by two scalar equations,

$$\phi_1(\boldsymbol{x},t) = 0, \qquad \phi_2(\boldsymbol{x},t) = 0. \tag{2.168}$$

Both of these constraints read in explicit form

$$\boldsymbol{x} = \boldsymbol{x}(y,t). \tag{2.169}$$

The number of degrees of freedom of a constrained single point is uniquely determined by its number of constraints. For the point bound to the three-dimensional path, we obtain $f = 3 - 2 = 1$ degrees of freedom.

Example 2.8 (Pendulum). A pendulum of time-varying length $L(t)$ can move on a spherical surface with a varying radius. This introduces a constraint which can be written in Cartesian coordinates as

$$\phi = r_1^2 + r_2^2 + r_3^2 - L^2(t) = 0 \tag{2.170}$$

or using (2.14), (2.15), and Fig. 2.2 in spherical coordinates as

$$\phi = |\boldsymbol{r}| - L(t) = 0. \tag{2.171}$$

A constraint with the Cartesian coordinates r_1, r_2 as generalized coordinates is written in explicit form

$$\boldsymbol{r}(r_1, r_2, t) = \begin{bmatrix} r_1 \\ r_2 \\ \pm\sqrt{L^2(t) - r_1^2 - r_2^2} \end{bmatrix} \tag{2.172}$$

or with the spherical coordinates ψ, ϑ as generalized coordinates

$$\boldsymbol{r}(\psi,\vartheta,t) = \begin{bmatrix} \cos\psi\sin\vartheta \\ \sin\psi\sin\vartheta \\ \cos\vartheta \end{bmatrix} L(t). \tag{2.173}$$

Often, curvilinear coordinates serve better for the introduction of constraints than Cartesian coordinates.

End of Example 2.8.

The constraints limit motion not only of single points in space but in particular the freedom of movement between several mass points of a mass point system. The number of degrees of freedom in a system of p points with q constraints is

$$f = 3p - q. \tag{2.174}$$

The q constraints can be described implicitly by an algebraic, generally nonlinear $q \times 1$ vector equation

$$\boldsymbol{\phi}(\boldsymbol{x},t) = \boldsymbol{0} \tag{2.175}$$

or explicitly by the $3p \times 1$ vector equation

$$\boldsymbol{x} = \boldsymbol{x}(\boldsymbol{y},t), \tag{2.176}$$

using the $f \times 1$ position vector of the constrained mass point system

$$\boldsymbol{y}(t) = \begin{bmatrix} y_1 \ y_2 \ \dots \ y_f \end{bmatrix}. \tag{2.177}$$

Constraints of the form (2.175) or (2.176), which simultaneously restrict the position and velocity of the system, are called geometric constraints. Another type of constraint is the integrable kinematic constraint of the form

$$\boldsymbol{\phi}(\boldsymbol{x},\dot{\boldsymbol{x}},t) = \boldsymbol{0}, \tag{2.178}$$

which does indeed depend formally on the velocity parameters, but can be converted to the form (2.175) via integration. Holonomic constraints comprise geometric and integrable kinematic constraints and can always be written in the form (2.175).

Time-invariant constraints are called scleronomic constraints, while time-variant constraints are designated as rheonomic constraints. In addition to the bilateral constraints characterized by Eq. (2.175), there are also unilateral constraints that lead to inequalities. In the form (2.176), unilateral constraints lead to a variable number of degrees of freedom, e.g. such as those that arise in contact problems. An extensive treatment of this subject can be found in Pfeiffer and Glocker [42].

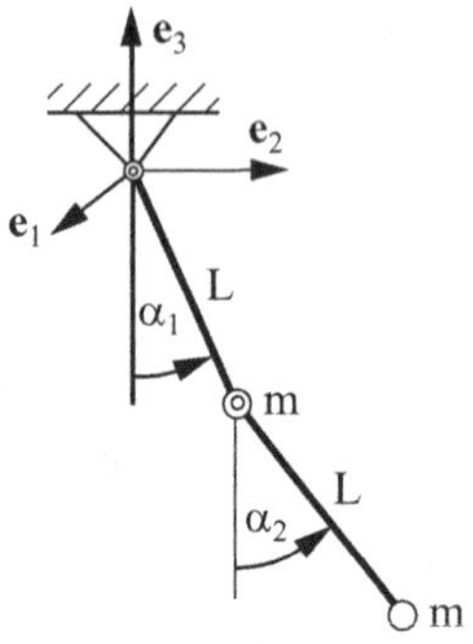

Fig. 2.14 Planar double pendulum

Example 2.9 (Planar Double Pendulum). The double pendulum, see Fig. 2.14, is a two-mass point system with four constraints (both points on the plane, both rod lengths are constant) and thus two degrees of freedom. The corresponding numbers are $p = 2, q = 4, f = 3p - q = 3 \cdot 2 - 4 = 2$. For the Cartesian coordinates

$$\boldsymbol{x}(t) = \begin{bmatrix} r_{11} \; r_{12} \; r_{13} \; r_{21} \; r_{22} \; r_{23} \end{bmatrix} \tag{2.179}$$

and the angular coordinates

$$\boldsymbol{y}(t) = \begin{bmatrix} \alpha_1 \; \alpha_2 \end{bmatrix} \tag{2.180}$$

the scleronomic constraints are written in implicit form

$$\boldsymbol{\phi} = \begin{bmatrix} r_{11} \\ r_{12}^2 + r_{13}^2 - L^2 \\ r_{21} \\ (r_{22} - r_{12})^2 + (r_{23} - r_{13})^2 - L^2 \end{bmatrix} = \boldsymbol{0} \tag{2.181}$$

and in explicit form

$$\boldsymbol{x} = \begin{bmatrix} 0 \\ L\sin\alpha_1 \\ -L\cos\alpha_1 \\ 0 \\ L\sin\alpha_1 + L\sin\alpha_2 \\ -L\cos\alpha_1 - L\cos\alpha_2 \end{bmatrix}. \tag{2.182}$$

We can confirm the equivalence of both forms by inserting (2.182) into (2.181).

End of Example 2.9.

The translation of a holonomic mass point system is obtained from (2.24) and (2.176), yielding

$$\boldsymbol{r}_i(t) = \boldsymbol{r}_i(\boldsymbol{y}, t), \qquad i = 1(1)p. \tag{2.183}$$

For the velocity we obtain

$$\mathbf{v}_i(t) = \frac{\partial \mathbf{r}_i}{\partial \mathbf{y}} \cdot \dot{\mathbf{y}}(t) + \frac{\partial \mathbf{r}_i}{\partial t} = \mathbf{J}_{Ti}(\mathbf{y},t) \cdot \dot{\mathbf{y}}(t) + \bar{\mathbf{v}}_i(\mathbf{y},t), \qquad i = 1(1)p, \tag{2.184}$$

where, besides the $3 \times f$ Jacobian matrix $\mathbf{J}_{Ti}$ of translation, the local 3×1 velocity vector $\bar{\mathbf{v}}_i$ can appear in the case of rheonomic constraints. For the acceleration one obtains likewise

$$\begin{aligned} \mathbf{a}_i(t) &= \mathbf{J}_{Ti}(\mathbf{y},t) \cdot \ddot{\mathbf{y}}(t) + \dot{\mathbf{J}}_{Ti}(\mathbf{y},t) \cdot \dot{\mathbf{y}}(t) + \frac{d\bar{\mathbf{v}}_i}{dt} \\ &= \mathbf{J}_{Ti}(\mathbf{y},t) \cdot \ddot{\mathbf{y}}(t) + \bar{\mathbf{a}}_i(\mathbf{y},\dot{\mathbf{y}},t), \qquad i = 1(1)p. \end{aligned} \tag{2.185}$$

In the scleronomic case, the 3×1 acceleration vector $\bar{\mathbf{a}}_i$ is quadratically dependent on the first derivative of the position vector. With rheonomic constraints on the other hand, terms can also arise that, in purely mechanical systems, depend either linearly or not at all on the first derivative $\dot{\mathbf{y}}(t)$ of the position vector. These terms are calculated with the help of (2.185).

In addition to the real motions of a system, virtual motions are also important in dynamics. A virtual motion is an arbitrary, infinitesimal motion of the system which is compatible with scleronomic and rheonomic constraints (provided they are "frozen" at the given point in time). The symbol δ of virtual quantities possesses the properties of variations in mathematics. The following applies for holonomic constraints

$$\begin{aligned} &\delta\mathbf{r} \neq \mathbf{0} \quad \text{for movable bearings,} \\ &\delta\mathbf{r} = \mathbf{0} \quad \text{for firm restraints,} \\ &\delta t = 0. \end{aligned} \tag{2.186}$$

The virtual motion of a point is thus determined by the virtual displacement $\delta\mathbf{r}$, while time is not varied. With virtual motions, calculations are made the same way as with differentials

$$\delta(c\mathbf{r}) = c\delta\mathbf{r}, \qquad \delta(\mathbf{r}_1 + \mathbf{r}_2) = \delta\mathbf{r}_1 + \delta\mathbf{r}_2, \qquad \delta\mathbf{r}(\mathbf{y}) = \frac{\partial \mathbf{r}}{\partial \mathbf{y}} \cdot \delta\mathbf{y}. \tag{2.187}$$

The following applies in particular for the virtual motion of the ith point

$$\delta\mathbf{r}_i = \mathbf{J}_{Ti} \cdot \delta\mathbf{y}, \qquad i = 1(1)p. \tag{2.188}$$

The virtual change of position $\delta\mathbf{y}$ determines, using the Jacobian matrices $\mathbf{J}_{Ti}$, the entire virtual motion of the system.

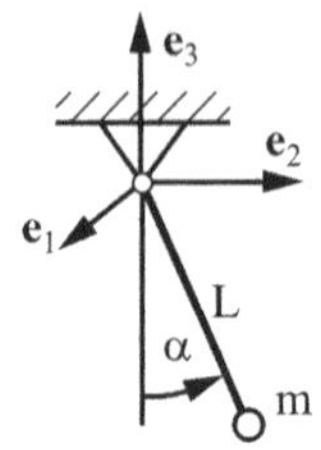

Fig. 2.15 Mathematical pendulum

According to the chain rule (2.11), there is a close connection between the Jacobian matrices $\boldsymbol{H}_{Ti}$ of the free system and $\boldsymbol{J}_{Ti}$ of the constrained system. With (2.8), (2.176) and (2.183), the following relation applies in particular,

$$\boldsymbol{J}_{Ti} = \frac{\partial \boldsymbol{r}_i}{\partial \boldsymbol{y}} = \frac{\partial \boldsymbol{r}_i}{\partial \boldsymbol{x}} \cdot \frac{\partial \boldsymbol{x}}{\partial \boldsymbol{y}} = \boldsymbol{H}_{Ti}(\boldsymbol{y},t) \cdot \boldsymbol{I}(\boldsymbol{y},t) \tag{2.189}$$

with the $3p \times f$ matrix $\boldsymbol{I}(\boldsymbol{y},t)$. In this way, the practical calculation of the Jacobian matrices can often be simplified considerably.

Example 2.10 (Mathematical Pendulum). The mathematical pendulum is a planar pendulum with one degree of freedom, see Fig. 2.15. With spherical coordinates as generalized coordinates, see (2.14), the constraint equation is

$$\boldsymbol{x} = \left[\frac{\pi}{2} \; (\pi - \alpha) \; L \right]. \tag{2.190}$$

With this we obtain

$$\frac{\partial \boldsymbol{x}}{\partial \alpha} = \left[0 \; -1 \; 0 \right]. \tag{2.191}$$

Taking (2.16) into account, from (2.189) we thus obtain the 3×1 Jacobian matrix

$$\boldsymbol{J}_T = \frac{\partial \boldsymbol{r}}{\partial \boldsymbol{x}} \cdot \frac{\partial \boldsymbol{x}}{\partial \alpha} = \begin{bmatrix} 0 \\ L\cos\alpha \\ L\sin\alpha \end{bmatrix}. \tag{2.192}$$

This result can easily be checked by means of direct partial differentiation of the location vector (2.193), *End of Example 2.10*,

$$\boldsymbol{r}(\alpha) = \begin{bmatrix} 0 \\ L\sin\alpha \\ -L\cos\alpha \end{bmatrix}. \tag{2.193}$$

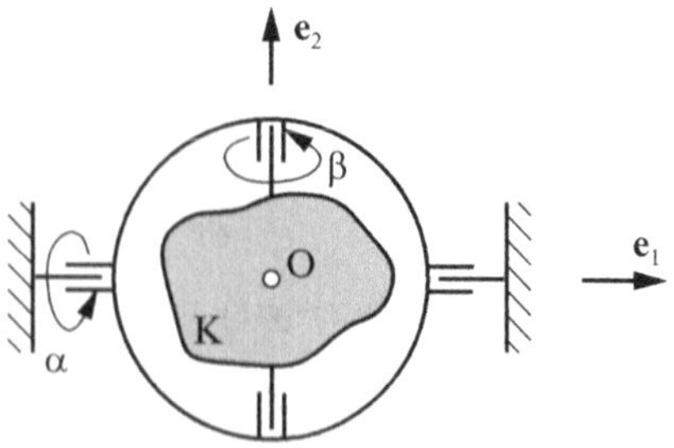

Fig. 2.16 Rotation of a rigid body in the Cardano joint

2.2.2 Multibody Systems

Just like the translation of a point, the rotation of a rigid body can also be restricted. The rotation of a rigid body K in a Cardano joint, see Fig. 2.16, is uniquely described by two degrees of freedom with the Cardano angles $\alpha(t), \beta(t)$ as generalized coordinates,

$$\boldsymbol{S}(t) = \boldsymbol{S}(\alpha, \beta). \tag{2.194}$$

The associated constraint is written implicitly with (2.99)

$$\phi(\boldsymbol{x}) = \gamma - \gamma_0 = 0 \tag{2.195}$$

and explicitly

$$\boldsymbol{x} = \boldsymbol{x}(\alpha, \beta) = \begin{bmatrix} \alpha \ \beta \ \gamma_0 \end{bmatrix}. \tag{2.196}$$

We see that the relations (2.165) and (2.166) found for the translation of a point can be transferred directly to the rotation of a body.

The number of degrees of freedom in a system of p rigid bodies with q constraints is

$$f = 6p - q. \tag{2.197}$$

For the q constraints, (2.175)–(2.177) are valid again, whereby (2.176) represents, in the case of a multibody system, a $6p \times 1$ vector equation.

The position and orientation of a holonomic multibody system is described by

$$\boldsymbol{r}_i(t) = \boldsymbol{r}_i(\boldsymbol{y}, t), \qquad \boldsymbol{S}_i(t) = \boldsymbol{S}_i(\boldsymbol{y}, t), \qquad i = 1(1)p \tag{2.198}$$

in accordance with (2.122) and (2.176). Supplementing (2.184) and (2.185), the following applies for the rotation,

$$\boldsymbol{\omega}_i(t) = \frac{\partial \boldsymbol{s}_i}{\partial \boldsymbol{y}} \cdot \dot{\boldsymbol{y}}(t) + \frac{\partial \boldsymbol{s}_i}{\partial t} = \boldsymbol{J}_{Ri}(\boldsymbol{y},t) \cdot \dot{\boldsymbol{y}}(t) + \overline{\boldsymbol{\omega}}_i(\boldsymbol{y},t), \quad i = 1(1)p, \tag{2.199}$$

$$\begin{aligned}\boldsymbol{\alpha}_i(t) &= \boldsymbol{J}_{Ri}(\boldsymbol{y},t) \cdot \ddot{\boldsymbol{y}}(t) + \dot{\boldsymbol{J}}_{Ri}(\boldsymbol{y},t) \cdot \dot{\boldsymbol{y}} + \dot{\overline{\boldsymbol{\omega}}}_i(\boldsymbol{y},t) \\ &= \boldsymbol{J}_{Ri}(\boldsymbol{y},t) \cdot \ddot{\boldsymbol{y}}(t) + \overline{\boldsymbol{\alpha}}_i(\boldsymbol{y},\dot{\boldsymbol{y}},t), \qquad i = 1(1)p.\end{aligned} \tag{2.200}$$

Here, the instantaneous infinitesimal 3×1 rotation vector $\boldsymbol{s}_i$ from (2.85) has been used again, and like before the comments regarding (2.113) and (2.114) are applicable for the calculation of the $3 \times f$ Jacobian matrix $\boldsymbol{J}_{Ri}$ of rotation. Also, $\overline{\boldsymbol{\omega}}_i$ is the local 3×1 rotation velocity vector and $\overline{\boldsymbol{\alpha}}_i$ is a 3×1 local rotation acceleration vector defined by means of (2.185).

We obtain for the virtual motion of the multibody system

$$\delta \boldsymbol{r}_i = \boldsymbol{J}_{Ti} \cdot \delta \boldsymbol{y}, \qquad \delta \boldsymbol{s}_i = \boldsymbol{J}_{Ri} \cdot \delta \boldsymbol{y}, \qquad i = 1(1)p, \tag{2.201}$$

supplementing (2.188). Also, the following applies in accordance with (2.189),

$$\boldsymbol{J}_{Ri} = \boldsymbol{H}_{Ri}(\boldsymbol{y},t) \cdot \boldsymbol{I}(\boldsymbol{y},t), \tag{2.202}$$

a relation that is very valuable for calculating the Jacobian matrix of rotation.

Example 2.11 (Cardano Joint). The Cardano point, see Fig. 2.16, is a two-body system with ten constraints and two degrees of freedom, $p = 2, q = 10, f = 6p - q = 6 \cdot 2 - 10 = 2$. For the 12×1 position vector of the free system

$$\boldsymbol{x}(t) = [r_{11} \; r_{12} \; r_{13} \; r_{21} \; r_{22} \; r_{23} \; \alpha_1 \; \beta_1 \; \gamma_1 \; \alpha_2 \; \beta_2 \; \gamma_2] \tag{2.203}$$

and the 2×1 position vector

$$\boldsymbol{y}(t) = [\alpha \; \beta] \tag{2.204}$$

the explicit constraints are

$$\boldsymbol{x} = [0\,0\,0\,0\,0\,0\,\alpha\,\beta\,0\,\alpha\,0\,0]. \tag{2.205}$$

Here we took into account the fact that the origin O of the frame is a fixed point of both bodies. Taking (2.100) and (2.202) into consideration, we obtain for the Jacobian matrices

$$\boldsymbol{J}_{T1} = \boldsymbol{J}_{T2} = \boldsymbol{0}, \quad \boldsymbol{J}_{R1} = \begin{bmatrix} 1 & 0 \\ 0 & \cos\alpha \\ 0 & \sin\alpha \end{bmatrix}, \quad \boldsymbol{J}_{R2} = \begin{bmatrix} 1 & 0 \\ 0 & 0 \\ 0 & 0 \end{bmatrix} \tag{2.206}$$

and the accelerations are

$$\boldsymbol{a}_1(t) = \boldsymbol{a}_2(t) = \mathbf{0}, \tag{2.207}$$

$$\boldsymbol{\alpha}_1(t) = \begin{bmatrix} \ddot{\alpha} \\ \ddot{\beta}\cos\alpha - \dot{\alpha}\dot{\beta}\sin\alpha \\ \ddot{\beta}\sin\alpha + \dot{\alpha}\dot{\beta}\cos\alpha \end{bmatrix}, \qquad \boldsymbol{\alpha}_2(t) = \begin{bmatrix} \ddot{\alpha} \\ 0 \\ 0 \end{bmatrix}. \tag{2.208}$$

In practice, we dispense with writing the $6p \times 1$ position vector $\boldsymbol{x}(t)$ of the free system in the case of large multibody systems, since this leads to long expressions, as (2.203) shows. The relations (2.198) are then evaluated directly with the $f \times 1$ position vector $\mathbf{y}(t)$.

End of Example 2.11.

By definition, holonomic systems also include free systems as a special case. Specifically, $q = 0, f = 6p$, $\mathbf{x} = \mathbf{y}$, $\boldsymbol{H}_{Ti} = \boldsymbol{J}_{Ti}$, $\boldsymbol{H}_{Ri} = \boldsymbol{J}_{Ri}$, $\boldsymbol{I} = \boldsymbol{E}$ then applies, i.e. the functional matrix $\boldsymbol{I}$ becomes the $6p \times 6p$ unit matrix.

2.2.3 Continuum

Constraints in a continuum are of a more theoretical nature since they cannot be influenced constructively. Nevertheless, we can easily model highly varying stiffness properties with good approximation by means of constraints. It is then possible to go from a general three-dimensional problem to a simpler two or one-dimensional task. One typical example for a holonomic constraint in a continuum is the Bernoulli hypothesis of beam bending, which requires planar cross-sectional surfaces – even under stress.

Deformation of a continuum is generally dependent on location, so the constraints must also be formulated locally. The constraints are then given as functions of the deformation gradient

$$\boldsymbol{\phi}(\boldsymbol{F}(\boldsymbol{\rho},t)) = \mathbf{0}. \tag{2.209}$$

Rigidity is a constraint typical for a continuum. With (2.32), we can write

$$\boldsymbol{\phi} = \boldsymbol{F}^T \cdot \boldsymbol{F} - \boldsymbol{E} = \mathbf{0}, \tag{2.210}$$

subjecting the nine coordinates of the deformation gradient to six constraints, so the three degrees of freedom of rotation remain. Besides the internal constraints given by (2.209), a continuum can also be bound to its surroundings. Additional external constraints then arise,

$$\boldsymbol{\phi}(\boldsymbol{r}(\boldsymbol{\rho},t)) = \mathbf{0} \qquad \text{on } A^r, \tag{2.211}$$

which correspond to the boundary conditions on the surface A^r. Boundary conditions restrict deformation on a plane or line segment or at discrete single points on the surface.

Example 2.12 (Torsion of a Round Bar). A twisted round bar with the current configuration (2.128) represents a continuum characterized by six degrees of freedom of rigid body motion and infinitely many degrees of freedom of torsion. In particular, we obtain from (2.130) the relations

$$\phi_1 = U_{11}^2 - 1 = 0, \tag{2.212}$$

$$\phi_2 = U_{22}^2 - 1 = 0, \tag{2.213}$$

$$\phi_3 = U_{33}^2 - 1 = 0, \tag{2.214}$$

$$\phi_4 = U_{23}^2 = 0, \tag{2.215}$$

$$\phi_5 = \rho_2 U_{12}^2 + \rho_3 U_{13}^2 = 0. \tag{2.216}$$

These constraints express the fact that the cross-sectional areas remain flat and undistorted under stress. Also, the round bar can be attached at three points on its left end. The external constraints are then

$$\begin{aligned}
&r_1 - \rho_1 = 0,\ r_2 - \rho_2 = 0,\ r_3 - \rho_3 = 0, && \text{for } \boldsymbol{\rho} = [0\ 0\ \tfrac{R}{2}],\\
&r_1 - \rho_1 = 0,\ r_2 - \rho_2 = 0, && \text{for } \boldsymbol{\rho} = [0\ 0\ -\tfrac{R}{2}],\\
&r_1 - \rho_1 = 0, && \text{for } \boldsymbol{\rho} = [0\ \tfrac{R}{2}\ 0].
\end{aligned} \tag{2.217}$$

The number of degrees of freedom is $f \to \infty$ for the one-dimensional problem of torsion.

End of Example 2.12.

2.3 Nonholonomic Systems

While holonomic constraints limit the freedom of motion of the position variables, and thus simultaneously of the velocity variables, nonholonomic constraints lead only to a restriction of the velocity, not of the position. Nonholonomic constraints are relatively rare in technology. Linear nonholonomic constraints can be realized purely mechanically, e.g. via rolling rigid wheels, while nonlinear nonholonomic constraints require the use of control devices. However, with the help of nonholonomic constraints, generalized velocities can be employed for a simplified description of holonomic systems as well.

Fig. 2.17 Rolling ball

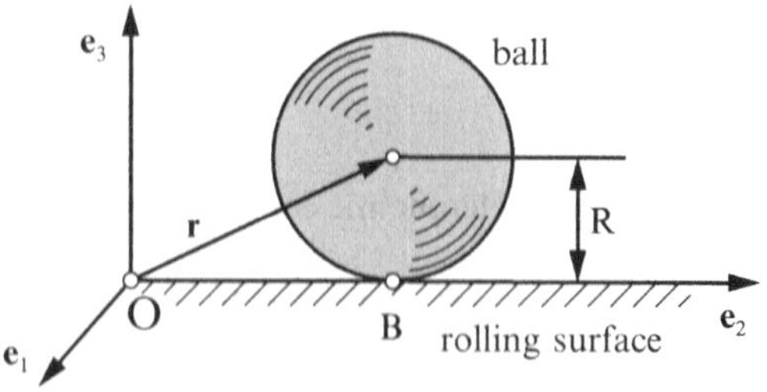

The number f of degrees of freedom of the position of a holonomic system is reduced by r nonholonomic constraints to the number g of degrees of freedom of velocity. Thus the following applies for a system of p rigid bodies,

$$g = f - r = 6p - q - r, \tag{2.218}$$

where (2.197) has been taken into account. The r nonholonomic constraints can be represented implicitly by the non-integrable $r \times 1$ vector equation

$$\boldsymbol{\psi}(\mathbf{y}, \dot{\mathbf{y}}, t) = \mathbf{0} \tag{2.219}$$

or explicitly by the $f \times 1$ vector differential equation

$$\dot{\mathbf{y}} = \dot{\mathbf{y}}(\mathbf{y}, \mathbf{z}, t), \tag{2.220}$$

yielding the $g \times 1$ vector of the generalized velocity coordinates

$$\mathbf{z}(t) = \begin{bmatrix} z_1 \; z_2 \; \dots \; z_g \end{bmatrix}. \tag{2.221}$$

The nonholonomic constraints belong to the kinematic constraints, and they can be scleronomic or rheonomic. It is however an essential condition that (2.219) cannot be integrated. Otherwise the constraints will be holonomic, see (2.175).

Example 2.13 (Rolling Ball). A ball (radius R) rolling on a surface, Fig. 2.17, is a rigid body with a holonomic (motion on a plane) and two nonholonomic constraints (rolling without slipping), $p = 1$, $q = 1$, $r = 2$, $f = 5$, $g = 3$. With the generalized coordinates of the free ball

$$\boldsymbol{x}(t) = \begin{bmatrix} r_1 \; r_2 \; r_3 \; \alpha \; \beta \; \gamma \end{bmatrix}, \tag{2.222}$$

the generalized coordinates of the ball bound to the surface

$$\mathbf{y}(t) = \begin{bmatrix} r_1 \; r_2 \; \alpha \; \beta \; \gamma, \end{bmatrix}, \tag{2.223}$$

and the generalized velocities

$$z(t) = \begin{bmatrix} \omega_1 \ \omega_2 \ \omega_3 \end{bmatrix} \tag{2.224}$$

the holonomic, scleronomic constraint is

$$\phi = r_3 - R = 0 \quad \text{or} \quad \boldsymbol{x} = \begin{bmatrix} r_1 \ r_2 \ R \ \alpha \ \beta \ \gamma \end{bmatrix}. \tag{2.225}$$

The nonholonomic, scleronomic constraints are obtained implicitly from the rolling condition, yielding

$$\boldsymbol{\psi} = \begin{bmatrix} \dot{r}_1 - R(\dot{\beta}\cos\alpha - \dot{\gamma}\sin\alpha\cos\beta) \\ \dot{r}_2 + R(\dot{\alpha} + \dot{\gamma}\sin\beta) \end{bmatrix} = \mathbf{0} \tag{2.226}$$

and explicitly as

$$\dot{\boldsymbol{y}} = \begin{bmatrix} \omega_2 R \\ -\omega_1 R \\ \omega_1 + \omega_2 \sin\alpha\tan\beta - \omega_3\cos\alpha\tan\beta \\ \omega_2\cos\alpha + \omega_3\sin\alpha \\ -\omega_2\frac{\sin\alpha}{\cos\beta} + \omega_3\frac{\cos\alpha}{\cos\beta} \end{bmatrix}. \tag{2.227}$$

The fact was taken into account that the absolute velocity of the point of contact B disappears, and (2.100) and Table 2.2 were consulted.

End of Example 2.13.

The configuration of a nonholonomic multibody system is given in unchanged form by (2.198). The state of velocity on the other hand is obtained from (2.184), (2.199), and (2.220), yielding

$$\boldsymbol{v}_i = \boldsymbol{v}_i(\boldsymbol{y},\boldsymbol{z},t), \qquad \boldsymbol{\omega}_i = \boldsymbol{\omega}_i(\boldsymbol{y},\boldsymbol{z},t), \qquad i = 1(1)p. \tag{2.228}$$

We thereby obtain for the acceleration

$$\boldsymbol{a}_i(t) = \frac{\partial \boldsymbol{v}_i}{\partial \boldsymbol{z}}\cdot\dot{\boldsymbol{z}}(t) + \frac{\partial \boldsymbol{v}_i}{\partial \boldsymbol{y}}\cdot\dot{\boldsymbol{y}}(t) + \frac{\partial \boldsymbol{v}_i}{\partial t} = \boldsymbol{L}_{Ti}(\boldsymbol{y},\boldsymbol{z},t)\cdot\dot{\boldsymbol{z}}(t) + \overline{\boldsymbol{a}}_i(\boldsymbol{y},\boldsymbol{z},t) \tag{2.229}$$

and also

$$\boldsymbol{\alpha}_i(t) = \frac{\partial \boldsymbol{\omega}_i}{\partial \boldsymbol{z}}\cdot\dot{\boldsymbol{z}}(t) + \frac{\partial \boldsymbol{\omega}_i}{\partial \boldsymbol{y}}\cdot\dot{\boldsymbol{y}}(t) + \frac{\partial \boldsymbol{\omega}_i}{\partial t} = \boldsymbol{L}_{Ri}(\boldsymbol{y},\boldsymbol{z},t)\cdot\dot{\boldsymbol{z}}(t) + \overline{\boldsymbol{\alpha}}_i(\boldsymbol{y},\boldsymbol{z},t). \tag{2.230}$$

For brevity's sake, the $3 \times g$ Jacobian matrices $\boldsymbol{L}_{Ti}$ and $\boldsymbol{L}_{Ri}$ and the local 3×1 acceleration vectors $\overline{\boldsymbol{a}}_i$ and $\overline{\boldsymbol{\alpha}}_i$ have been introduced as in the holonomic case.

In analogy to the virtual motions of holonomic systems, we can also introduce the virtual velocity of nonholonomic systems. A virtual velocity is an arbitrary,

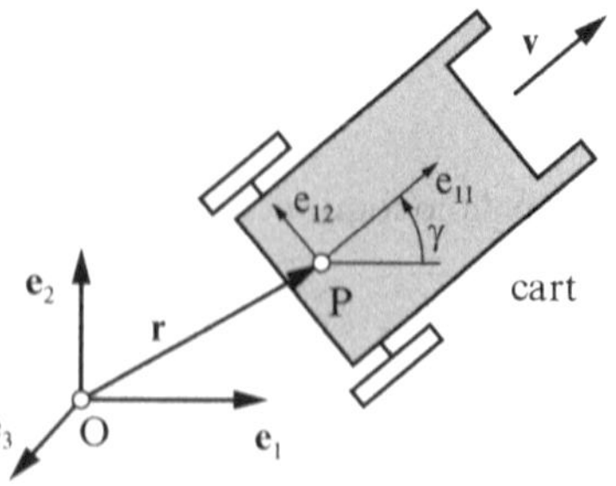

Fig. 2.18 Motion of a cart with rigid wheels

infinitesimal change in velocity which always agrees with the constraints. The symbol δ' of the virtual velocity has the properties

$$\delta' \boldsymbol{r}_i = \delta' \boldsymbol{s}_i = \mathbf{0}, \qquad \delta' \boldsymbol{v}_i \neq \mathbf{0}, \qquad \delta' \boldsymbol{\omega}_i \neq \mathbf{0}, \qquad \delta' t = 0. \tag{2.231}$$

Thus, when determining the virtual velocity, the position and time are not varied. In particular, the following applies for the virtual velocity of a multibody system,

$$\delta' \boldsymbol{v}_i = \boldsymbol{L}_{Ti} \cdot \delta' \boldsymbol{z}, \qquad \delta' \boldsymbol{\omega}_i = \boldsymbol{L}_{Ri} \cdot \delta' \boldsymbol{z}, \qquad i = 1(1)p. \tag{2.232}$$

The virtual change in velocity $\delta' \boldsymbol{z}$ determines, via the functional matrices $\boldsymbol{L}_{Ti}$, $\boldsymbol{L}_{Ri}$, the total virtual velocity of the system.

Also, there is a close connection between the different Jacobian matrices, as was already clarified by (2.202). The following is true,

$$\boldsymbol{L}_{Ti}(\boldsymbol{y},\boldsymbol{z},t) = \boldsymbol{J}_{Ti}(\boldsymbol{y},t) \cdot \boldsymbol{K}(\boldsymbol{y},\boldsymbol{z},t) \tag{2.233}$$

$$\boldsymbol{L}_{Ri}(\boldsymbol{y},\boldsymbol{z},t) = \boldsymbol{J}_{Ri}(\boldsymbol{y},t) \cdot \boldsymbol{K}(\boldsymbol{y},\boldsymbol{z},t) \tag{2.234}$$

with the $f \times g$ matrix

$$\boldsymbol{K}(\boldsymbol{y},\boldsymbol{z},t) = \frac{\partial \dot{\boldsymbol{y}}(\boldsymbol{y},\boldsymbol{z},t)}{\partial \boldsymbol{z}}. \tag{2.235}$$

With this we can often simplify our calculation of the Jacobian matrices.

Example 2.14 (Transport Cart). A transport cart with two independent, massless wheels, see Fig. 2.18, is characterized by the fact that the axial center P cannot move in the body-fixed 2-direction as a result of the static friction forces of the wheels. Assuming planar motion, we are concerned with a body with three holonomic constraints and one nonholonomic constraint, $p = 1, q = 3, r = 1, f = 3, g = 2$. With the 6×1 position vector of the free body

$$\boldsymbol{x}(t) = \begin{bmatrix} r_1 \ r_2 \ r_3 \ \alpha \ \beta \ \gamma \end{bmatrix}, \tag{2.236}$$

the 3×1 position vector of the cart

$$\boldsymbol{y}(t) = \begin{bmatrix} r_1 \ r_2 \ \gamma \end{bmatrix} \tag{2.237}$$

and its 2×1 velocity vector

$$z(t) = \begin{bmatrix} v & \dot{\gamma} \end{bmatrix}, \tag{2.238}$$

the nonholonomic constraint is written in explicit form

$$\dot{y} = \begin{bmatrix} v\cos\gamma \\ v\sin\gamma \\ \dot{\gamma} \end{bmatrix}. \tag{2.239}$$

For the 3×3 Jacobian matrices we obtain

$$\boldsymbol{J}_T = \begin{bmatrix} 1 & 0 & 0 \\ 0 & 1 & 0 \\ 0 & 0 & 0 \end{bmatrix}, \qquad \boldsymbol{J}_R = \begin{bmatrix} 0 & 0 & 0 \\ 0 & 0 & 0 \\ 0 & 0 & 1 \end{bmatrix}, \tag{2.240}$$

the 3×2 functional matrix is written as

$$\boldsymbol{K}(\boldsymbol{y}) = \begin{bmatrix} \cos\gamma & 0 \\ \sin\gamma & 0 \\ 0 & 1 \end{bmatrix}, \tag{2.241}$$

with which we can also determine using (2.233), (2.234), the 3×2 functional matrices $\boldsymbol{L}_T, \boldsymbol{L}_R$ where

$$\boldsymbol{L}_T = \begin{bmatrix} \cos\gamma & 0 \\ \sin\gamma & 0 \\ 0 & 0 \end{bmatrix}, \qquad \boldsymbol{L}_R = \begin{bmatrix} 0 & 0 \\ 0 & 0 \\ 0 & 1 \end{bmatrix}. \tag{2.242}$$

We also find

$$\bar{\boldsymbol{a}} = \begin{bmatrix} -v\dot{\gamma}\sin\gamma \\ v\dot{\gamma}\cos\gamma \\ 0 \end{bmatrix}, \qquad \bar{\boldsymbol{\alpha}} = \boldsymbol{0}. \tag{2.243}$$

Using (2.229), (2.230), this also yields the state of acceleration.

End of Example 2.14.

The nonholonomic constraints (2.219) and (2.220) are sometimes also designated as first-class nonholonomic constraints in order to differentiate them from second-class holonomic constraints, see e.g. Hamel [25]. Second-class nonholonomic constraints restrict accelerations, which is however only of theoretical interest.

Nonholonomic systems also include all holonomic systems as a special case. Since the concept of generalized velocities is lacking in the case of holonomic systems, this special case is not trivial, since the following is then true: $r = 0$, $g = f$, $\boldsymbol{y} = \boldsymbol{y}(\boldsymbol{y}, \boldsymbol{z})$, $\boldsymbol{K} = \boldsymbol{K}(\boldsymbol{y}, \boldsymbol{z})$. It should also be mentioned that the case at hand (2.220) is always scleronomic and the $f \times f$ matrix $\boldsymbol{K}$ is usually regular and thus invertible. Generalized velocities offer especially for large holonomic multibody systems crucial advantages resulting from the separation of kinematics and dynamics.

Example 2.15 (Point Motion in Spherical Coordinates). The use of generalized velocities is already advantageous in the investigation of a simple point motion. The Jacobian matrix $\boldsymbol{H}_T$ can be reduced to a simpler functional matrix $\boldsymbol{L}_T$ by means of the appropriate choice of generalized velocities. With the generalized velocities

$$z(t) = \begin{bmatrix} r\dot{\psi} & r\dot{\vartheta} & \dot{r} \end{bmatrix} \tag{2.244}$$

we obtain

$$\dot{y}(y,z) = \begin{bmatrix} \frac{1}{r}(r\dot{\psi}) & \frac{1}{r}(r\dot{\vartheta}) & \dot{r} \end{bmatrix} \tag{2.245}$$

and the 3×3 matrix

$$\boldsymbol{K}(y) = \begin{bmatrix} \frac{1}{r} & 0 & 0 \\ 0 & \frac{1}{r} & 0 \\ 0 & 0 & 1 \end{bmatrix}, \tag{2.246}$$

which, together with (2.16) and (2.233), leads to a dimensionless regular 3×3 matrix $\boldsymbol{L}_T$. This moves the singularity to (2.246) – it cannot be avoided even with generalized velocities.

End of Example 2.15.

2.4 Relative Motion of the Frame

The previous discussions were always based on a spatially fixed frame that is not in motion. This assumption was especially important in the calculation of velocity and acceleration, see e.g. (2.5) and (2.12). Yet for many engineering problems, it is useful to introduce a moving frame in addition to the fixed frame. The motion of the frame can either be predefined as a target motion, or it is obtained directly as a particular solution from the equations of motion. In the neighborhood of the target motion or of a particular, periodic solution, we can then often execute a linearization of the motion.

2.4.1 Moving Frame

In addition to the spatially fixed inertial frame $\{0_I; \boldsymbol{e}_{I\alpha}\}$, now a moving reference frame $\{0_R; \boldsymbol{e}_{R\alpha}\}$, $\alpha = 1(1)3$ is introduced. The motion of the frame R is described with respect to the frame I by the 3×1 vector $\boldsymbol{r}_R(t)$ and the 3×3 rotation tensor $\boldsymbol{S}_R(t)$. For the basis vectors, the transformation law is applicable,

$$\boldsymbol{e}_{I\alpha} = \boldsymbol{S}_R(t) \cdot \boldsymbol{e}_{R\alpha}(t), \qquad \alpha = 1(1)3, \tag{2.247}$$

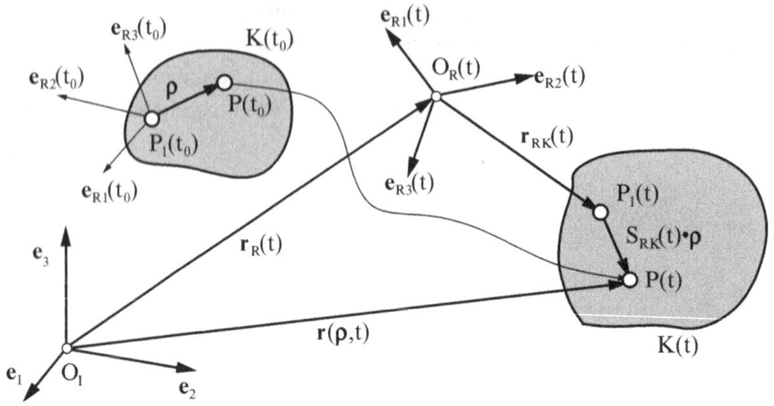

Fig. 2.19 Relative motion of a rigid body

which is analogously also valid for the coordinates of vectors and tensors. For the coordinates of a vector $\boldsymbol{a}$ or a tensor $\boldsymbol{A}$, we obtain the relation

$$_I\boldsymbol{a} = \boldsymbol{S}_R \cdot {_R\boldsymbol{a}}, \qquad _I\boldsymbol{A} = \boldsymbol{S}_R \cdot {_R\boldsymbol{A}} \cdot \boldsymbol{S}_R^T. \tag{2.248}$$

In case it is required, the frame is displayed by the lower left index.

Using Fig. 2.19, we thus obtain the following for the current configuration of a rigid body K,

$$_I\boldsymbol{r}(\boldsymbol{\rho},t) = {_I\boldsymbol{r}_R(t)} + \boldsymbol{S}_R(t) \cdot [{_R\boldsymbol{r}_{R1}(t)} + \boldsymbol{S}_{RK}(t) \cdot \boldsymbol{\rho}], \tag{2.249}$$

or completely written in the inertial frame I,

$$\boldsymbol{r}(\boldsymbol{\rho},t) = \boldsymbol{r}_R(t) + \boldsymbol{r}_{R1}(t) + \boldsymbol{S}_R(t) \cdot \boldsymbol{S}_{RK}(t) \cdot \boldsymbol{\rho}. \tag{2.250}$$

By comparing with (2.78), we thus obtain for the absolute position of the rigid body, expressed in the parameters of relative motion,

$$\boldsymbol{r}_1(t) = \boldsymbol{r}_R(t) + \boldsymbol{r}_{R1}(t), \tag{2.251}$$

$$\boldsymbol{S}(t) = \boldsymbol{S}_R(t) \cdot \boldsymbol{S}_{RK}(t). \tag{2.252}$$

If we also take into account the inverse deformation

$$\boldsymbol{\rho} = \boldsymbol{S}_{RK}^T(t) \cdot \boldsymbol{S}_R^T(t) \cdot \boldsymbol{r}_{RP}(\boldsymbol{\rho},t), \tag{2.253}$$

we then obtain, via the material derivative of (2.249), the absolute velocity

$$\begin{aligned} _I\boldsymbol{v}(\boldsymbol{\rho},t) &= \frac{d}{dt}{_I\boldsymbol{r}_R} + \frac{d}{dt}\boldsymbol{S}_R \cdot [{_R\boldsymbol{r}_{R1}} + \boldsymbol{S}_{RK} \cdot \boldsymbol{\rho}] + \boldsymbol{S}_R \cdot [\frac{d}{dt}{_R\boldsymbol{r}_{R1}} + \frac{d}{dt}\boldsymbol{S}_{RK} \cdot \boldsymbol{\rho}] \\ &= {_I\boldsymbol{r}_R^*} + {_I\tilde{\boldsymbol{\omega}}_R} \cdot {_I\boldsymbol{r}_{R1}} + {_I\dot{\boldsymbol{r}}_{R1}} + ({_I\tilde{\boldsymbol{\omega}}_R} + {_I\tilde{\boldsymbol{\omega}}_{RK}}) \cdot {_I\boldsymbol{r}_{RP}}, \end{aligned} \tag{2.254}$$

where $(^*)$ signifies the derivative in the inertial frame and $(\dot{})$ the derivative in the reference frame. Comparison with (2.86) yields the following for the laws of relative motion, see e.g. Magnus and Müller-Slany [36],

$$\boldsymbol{v}_1(t) = \boldsymbol{r}_R^*(t) + \tilde{\boldsymbol{\omega}}_R(t) \cdot \boldsymbol{r}_{R1}(t) + \dot{\boldsymbol{r}}_{RK}(t), \tag{2.255}$$

$$\boldsymbol{\omega}(t) = \boldsymbol{\omega}_R(t) + \boldsymbol{\omega}_{RK}(t). \tag{2.256}$$

A corresponding calculation finally leads to the following for the absolute acceleration of the relative motion,

$$\boldsymbol{a}_1(t) = \boldsymbol{r}_R^{**} + (\dot{\tilde{\boldsymbol{\omega}}}_R + \tilde{\boldsymbol{\omega}}_R \cdot \tilde{\boldsymbol{\omega}}_R) \cdot \boldsymbol{r}_{R1} + 2\tilde{\boldsymbol{\omega}}_R \cdot \dot{\boldsymbol{r}}_{R1} + \ddot{\boldsymbol{r}}_{R1}, \tag{2.257}$$

$$\boldsymbol{\alpha}(t) = \dot{\boldsymbol{\omega}}_R + \tilde{\boldsymbol{\omega}}_R \cdot \boldsymbol{\omega}_{RK} + \dot{\boldsymbol{\omega}}_{RK}. \tag{2.258}$$

In (2.257), the first two terms denote the guidance acceleration, the third term the Coriolis acceleration, and the fourth term the relative acceleration.

The reference frame R can also be attached to the rigid body K. We then call it a body-fixed frame $\{O_1, \boldsymbol{e}_{11}\}$. In this special case, the following applies,

$$\boldsymbol{r}_{R1}(t) = \boldsymbol{0}, \qquad \boldsymbol{S}_{RK}(t) = \boldsymbol{E} \tag{2.259}$$

and (2.250) turns directly into (2.78). This means that the motion of a rigid body can also be interpreted as the motion of a Cartesian frame that is connected to the rigid body. If we restrict ourselves to rigid body mechanics from the outset, this gives us an easy access to the kinematics. However, describing rigid body motion with body-fixed frames complicates the continuum-mechanical approach that is privileged in this book.

Given a multibody system, a separate reference frame $\{O_{jR}; \boldsymbol{e}_{jR\alpha}\}$, $\alpha = 1(1)3$, $j = 1(1)n$ can be selected for each partial body K_i, $i = 1(1)p$. Then

$$\boldsymbol{r}_i(t) = \boldsymbol{r}_{jR}(t) + \boldsymbol{r}_{jRi}(t), \tag{2.260}$$

$$\boldsymbol{S}_i(t) = \boldsymbol{S}_{jR}(t) \cdot \boldsymbol{S}_{jRi}(t) \tag{2.261}$$

applies and the relations (2.255)–(2.258) must also be generalized accordingly.

2.4.2 *Free and Holonomic Systems*

Holonomic systems include the free systems, $q = 0, f = 6p, \boldsymbol{x} = \boldsymbol{y}, \boldsymbol{I} = \boldsymbol{E}$, as a special case. Mass point systems represent a subgroup of multibody systems with $f = 3p$. For this reason, it will be sufficient to deal only with holonomic multibody systems in this context.

The number of degrees of freedom of a system is not changed by the introduction of one or more reference frames. The degrees of freedom can however be distributed varyingly to the reference and relative motions. If the reference motion is predefined by pure time functions, the relative motion encompasses all degrees of freedom. If we choose body-fixed reference frames exclusively on the other hand, all degrees of freedom are found in the reference motion. In the common case of a mixed distribution of degrees of freedom therefore

$$\boldsymbol{r}_R = \boldsymbol{r}_R(\boldsymbol{y},t), \qquad \boldsymbol{S}_R = \boldsymbol{S}_R(\boldsymbol{y},t) \tag{2.262}$$

applies. Assuming that all vectors and tensors are represented in the reference frame, we then obtain in accordance with (2.184), (2.199) for the guidance velocities of the reference motion

$$\boldsymbol{r}_R^* = \boldsymbol{S}_R^T \cdot \left(\frac{\partial \boldsymbol{S}_r \cdot \boldsymbol{r}_R}{\partial \boldsymbol{y}} \cdot \dot{\boldsymbol{y}} + \frac{\partial \boldsymbol{S}_r \cdot \boldsymbol{r}_R}{\partial t} \right) = \boldsymbol{J}_{TR}(\boldsymbol{y},t) \cdot \dot{\boldsymbol{y}}(t) + \bar{\boldsymbol{v}}_R(\boldsymbol{y},t), \tag{2.263}$$

$$\boldsymbol{\omega}_R = \frac{\partial \boldsymbol{s}_R}{\partial \boldsymbol{y}} \cdot \dot{\boldsymbol{y}} + \frac{\partial \boldsymbol{s}_R}{\partial t} = \boldsymbol{J}_{RR}(\boldsymbol{y},t) \cdot \dot{\boldsymbol{y}}(t) + \bar{\boldsymbol{\omega}}_R(\boldsymbol{y},t) \tag{2.264}$$

with the $3 \times f$ Jacobian matrices $\boldsymbol{J}_{TR}$ and $\boldsymbol{J}_{RR}$ of the guidance motion,

$$\boldsymbol{r}_{Ri} = \boldsymbol{r}_{Ri}(\boldsymbol{y},t), \qquad \boldsymbol{S}_{Ri} = \boldsymbol{S}_{Ri}(\boldsymbol{y},t) \tag{2.265}$$

and for the relative velocities we obtain likewise

$$\dot{\boldsymbol{r}}_{Ri} = \frac{\partial \boldsymbol{r}_{Ri}}{\partial \boldsymbol{y}} \cdot \dot{\boldsymbol{y}} + \frac{\partial \boldsymbol{r}_{Ri}}{\partial t} = \boldsymbol{J}_{TRi}(\boldsymbol{y},t) \cdot \dot{\boldsymbol{y}}(t) + \bar{\boldsymbol{v}}_{Ri}(\boldsymbol{y},t), \tag{2.266}$$

$$\boldsymbol{\omega}_{Ri} = \frac{\partial \boldsymbol{s}_{Ri}}{\partial \boldsymbol{y}} \cdot \dot{\boldsymbol{y}} + \frac{\partial \boldsymbol{s}_{Ri}}{\partial t} = \boldsymbol{J}_{RRi}(\boldsymbol{y},t) \cdot \dot{\boldsymbol{y}}(t) + \bar{\boldsymbol{\omega}}_{Ri}(y,t). \tag{2.267}$$

Here, $\boldsymbol{J}_{TRi}$ and $\boldsymbol{J}_{RRi}$ are the $3 \times f$ Jacobian matrices of the relative motion. We then find the accelerations for the guidance and relative motions following (2.185) and (2.200).

The absolute velocities and accelerations are then obtained with (2.262)–(2.267) from (2.255) to (2.258). We hereby find for the Jacobian matrices the relation

$$\boldsymbol{J}_{Ti} = \boldsymbol{J}_{TR} + \boldsymbol{J}_{TRi} - \tilde{\boldsymbol{r}}_{Ri} \cdot \boldsymbol{J}_{RR}, \tag{2.268}$$

$$\boldsymbol{J}_{Ri} = \boldsymbol{J}_{RR} + \boldsymbol{J}_{RRi}. \tag{2.269}$$

We can see that a purely time-dependent guidance motion does not at all affect the Jacobian matrices of the multibody system at hand, $\boldsymbol{J}_{TR} = \boldsymbol{J}_{RR} = \boldsymbol{0}$.

Example 2.16 (Overturning Double Pendulum). Both bodies of the double pendulum in Fig. 2.20 have a high initial velocity. The initial conditions are $\alpha_{10} = \alpha_{20} = 0$,

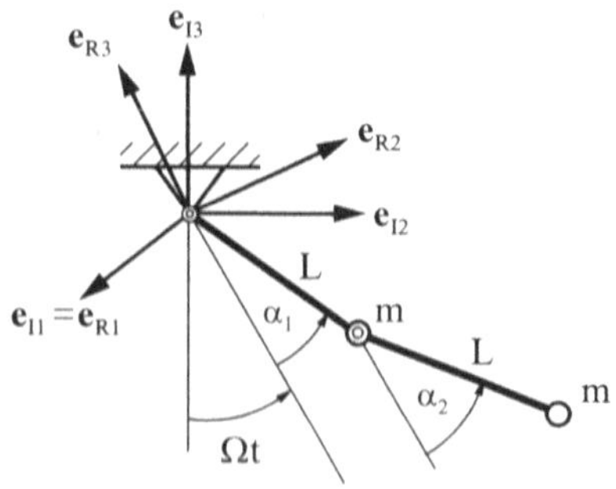

Fig. 2.20 Overturning double pendulum

$\dot{\alpha}_{10} = \dot{\alpha}_{20} = \Omega \gg \sqrt{g/L}$. In order to examine the motion, it is practical to use a reference frame that rotates with the rotation velocity Ω,

$$\boldsymbol{r}_R(t) = \boldsymbol{0}, \quad \boldsymbol{S}_R(t) = \begin{bmatrix} 1 & 0 & 0 \\ 0 & \cos\Omega t & -\sin\Omega t \\ 0 & \sin\Omega t & \cos\Omega t \end{bmatrix}, \tag{2.270}$$

$$\boldsymbol{r}_R^*(t) = \boldsymbol{0}, \quad \boldsymbol{\omega}_R = \overline{\boldsymbol{\omega}}_R = \begin{bmatrix} \Omega & 0 & 0 \end{bmatrix}, \quad \boldsymbol{J}_{TR} = \boldsymbol{J}_{RR} = \boldsymbol{0}. \tag{2.271}$$

Also, the following applies for the relative motion in the reference frame

$$\boldsymbol{r}_{R1} = \begin{bmatrix} 0 \\ \sin\alpha_1 \\ -\cos\alpha_1 \end{bmatrix} L, \qquad \boldsymbol{r}_{R2} = \begin{bmatrix} 0 \\ \sin\alpha_1 + \sin\alpha_2 \\ -\cos\alpha_1 - \cos\alpha_2 \end{bmatrix} L \tag{2.272}$$

with the Jacobian matrices

$$\boldsymbol{J}_{TR1} = \begin{bmatrix} 0 & 0 \\ \cos\alpha_1 & 0 \\ \sin\alpha_1 & 0 \end{bmatrix} L, \quad \boldsymbol{J}_{TR2} = \begin{bmatrix} 0 & 0 \\ \cos\alpha_1 & \cos\alpha_2 \\ \sin\alpha_1 & \sin\alpha_2 \end{bmatrix} L. \tag{2.273}$$

Observing (2.255), the absolute velocities in the reference frame read as

$$\boldsymbol{v}_1 = \begin{bmatrix} 0 \\ (\dot{\alpha}_1 + \Omega)\cos\alpha_1 \\ (\dot{\alpha}_1 + \Omega)\sin\alpha_1 \end{bmatrix} L, \tag{2.274}$$

$$\boldsymbol{v}_2 = \begin{bmatrix} 0 \\ (\dot{\alpha}_1 + \Omega)\cos\alpha_1 + (\dot{\alpha}_2 + \Omega)\cos\alpha_2 \\ (\dot{\alpha}_2 + \Omega)\sin\alpha_1 + (\dot{\alpha}_2 + \Omega)\sin\alpha_2 \end{bmatrix} L. \tag{2.275}$$

By means of the moving reference frame R, the Jacobian matrices remain in the simple form (2.273), even when using relative coordinates. Further advantages will be seen in the linearization of the motion.

End of Example 2.16.

2.4.3 Nonholonomic Systems

Nonholonomic constraints in the explicit form (2.220) can be inserted into the expressions (2.263)–(2.267) of the guidance and relative velocities, so then they depend too on the $g \times 1$ vector of the generalized velocities. The corresponding accelerations are again obtained using (2.229) and (2.230). We finally arrive at the absolute velocities and accelerations with (2.255)–(2.258). For the Jacobian matrices, we obtain the relation

$$\boldsymbol{L}_{Ti} = \boldsymbol{L}_{TR} + \boldsymbol{L}_{TRi} - \tilde{\boldsymbol{r}}_{Ri} \cdot \boldsymbol{L}_{RR}, \tag{2.276}$$

$$\boldsymbol{L}_{Ri} = \boldsymbol{L}_{RR} + \boldsymbol{L}_{RRi}. \tag{2.277}$$

It is also true here that a purely time-dependent guidance motion does not influence the Jacobian matrices of the multibody system under consideration, $\boldsymbol{L}_{TR} = \boldsymbol{L}_{RR} = \boldsymbol{0}$.

2.5 Linearization of the Kinematics

We have already discussed the linearization of kinematic relations in our discussion of continua in Sect. 2.1.3. In this section, the linearization of the motion of point and multibody systems with respect to an arbitrary target motion will be considered. We will again ignore the distinction between free and holonomic systems in this context.

In engineering, a target motion $\boldsymbol{y}_S(t)$ is often defined by the task of a machine or device, whereby the actual motion $\boldsymbol{y}(t)$ deviates only slightly from it. If it is true that the velocities $\dot{\boldsymbol{y}}(t)$ and accelerations $\ddot{\boldsymbol{y}}(t)$ essentially also correspond to the target motion, then the following is valid for holonomic systems,

$$\boldsymbol{y}(t) = \boldsymbol{y}_S(t) + \boldsymbol{\eta}(t), \qquad |\boldsymbol{\eta}(t)| \ll a, \tag{2.278}$$

$$\dot{\boldsymbol{y}}(t) = \dot{\boldsymbol{y}}_S(t) + \dot{\boldsymbol{\eta}}(t), \qquad |\dot{\boldsymbol{\eta}}(t)| \ll b, \tag{2.279}$$

$$\ddot{\boldsymbol{y}}(t) = \ddot{\boldsymbol{y}}_S(t) + \ddot{\boldsymbol{\eta}}(t), \qquad |\ddot{\boldsymbol{\eta}}(t)| \ll c, \tag{2.280}$$

where $\boldsymbol{\eta}(t)$ is the $f \times 1$ position vector of the small deviations and a, b, c represent suitable reference values. With (2.278), we obtain from (2.198) after a Taylor series expansion,

$$\boldsymbol{r}_i(\boldsymbol{\eta}, t) = \boldsymbol{r}_{iS}(t) + \boldsymbol{J}_{TiS}(t) \cdot \boldsymbol{\eta} + \boldsymbol{r}_{i2}(\boldsymbol{\eta} \cdot \boldsymbol{\eta}, t) + \ldots, \tag{2.281}$$

$$\boldsymbol{S}_i(\boldsymbol{\eta}, t) = \boldsymbol{S}_{iS}(t) + \boldsymbol{S}_{i1}(\boldsymbol{\eta}, t) + \boldsymbol{S}_{i2}(\boldsymbol{\eta} \cdot \boldsymbol{\eta}, t) + \ldots, \tag{2.282}$$

where $\boldsymbol{r}_{iS}(t)$ and $\boldsymbol{S}_{iS}(t)$ denote the 3×1 position vector and the 3×3 rotation tensor of the target motion. Also, the following applies according to (2.184) for the linearized $3 \times f$ Jacobian matrix of translation

$$\boldsymbol{J}_{Ti}(\boldsymbol{\eta},t) = \boldsymbol{J}_{TiS}(t) + \boldsymbol{J}_{Ti1}(\boldsymbol{\eta},t) + \ldots \quad . \tag{2.283}$$

For the linear term of the series expansion of the Jacobian matrix it follows

$$\boldsymbol{J}_{Ti1}(\boldsymbol{\eta},t) = \frac{\partial \boldsymbol{r}_{i2}(\boldsymbol{\eta}\cdot\boldsymbol{\eta},t)}{\partial \boldsymbol{\eta}}. \tag{2.284}$$

If we now neglect the quadratic and higher components, we obtain for the velocity and acceleration of holonomic systems

$$\boldsymbol{v}_i(t) = \boldsymbol{J}_{TiS}(t)\cdot\dot{\boldsymbol{\eta}}(t) + \dot{\boldsymbol{J}}_{TiS}(t)\cdot\boldsymbol{\eta}(t) + \boldsymbol{v}_{iS}(t), \tag{2.285}$$

$$\boldsymbol{a}_i(t) = \boldsymbol{J}_{TiS}(t)\cdot\ddot{\boldsymbol{\eta}}(t) + 2\dot{\boldsymbol{J}}_{TiS}(t)\cdot\dot{\boldsymbol{\eta}}(t) + \ddot{\boldsymbol{J}}_{TiS}(t)\cdot\boldsymbol{\eta}(t) + \boldsymbol{a}_{iS}(t), \tag{2.286}$$

while for the virtual translational motion

$$\delta\boldsymbol{r}_i = \left[\boldsymbol{J}_{TiS}(t) + \boldsymbol{J}_{Ti1}(\boldsymbol{\eta},t)\right]\cdot\delta\boldsymbol{\eta} \tag{2.287}$$

is true.

Equation (2.283) applies analogously for the linearized $3 \times f$ Jacobian matrix of rotation. Calculation of the Jacobian matrices $\boldsymbol{J}_{RiS}(t)$ and $\boldsymbol{J}_{Ri1}(t)$ is much more complicated however. Taking the definition found in (2.113), (2.114) into account, we obtain

$$\frac{\partial \tilde{s}_{iS\alpha\beta}}{\partial \eta_\delta} = \frac{\partial S_{i1\alpha\gamma}}{\partial \eta_\delta} S_{iS\beta\gamma}, \tag{2.288}$$

$$\frac{\partial \tilde{s}_{i1\alpha\beta}}{\partial \eta_\delta} = \frac{\partial S_{i2\alpha\gamma}}{\partial \eta_\delta} S_{iS\beta\gamma} + \frac{\partial S_{i1\alpha\gamma}}{\partial \eta_\delta} S_{i1\beta\gamma}, \qquad \alpha,\beta,\gamma = 1(1)3,\ \delta = 1(1)f. \tag{2.289}$$

The rotation speed and rotation acceleration are thus

$$\begin{aligned}\boldsymbol{\omega}_i(t) &= \boldsymbol{J}_{RiS}(t)\cdot\dot{\boldsymbol{\eta}}(t) + \boldsymbol{J}'_{RiS}(t)\cdot\boldsymbol{\eta}(t) + \boldsymbol{\omega}_{iS}(t),\\ \boldsymbol{J}'_{RiS}(t)\boldsymbol{\eta}(t) &= \frac{\partial \boldsymbol{S}_{iS}}{\partial t}\cdot\boldsymbol{S}_{i1}^T + \frac{\partial \boldsymbol{S}_{i1}}{\partial t}\cdot\boldsymbol{S}_{iS}^T,\end{aligned} \tag{2.290}$$

$$\boldsymbol{\alpha}_i(t) = \boldsymbol{J}_{RiS}(t)\cdot\ddot{\boldsymbol{\eta}}(t) + (\dot{\boldsymbol{J}}_{RiS}(t) + \boldsymbol{J}'_{RiS}(t))\cdot\dot{\boldsymbol{\eta}}(t) + \dot{\boldsymbol{J}}'_{RiS}(t)\cdot\boldsymbol{\eta}(t) + \boldsymbol{\alpha}_{iS}(t), \tag{2.291}$$

while the virtual rotation corresponds to the relation (2.287).

We can see that the rotation, because of its nonlinearity, involves considerably more complexity in the linearization process than the translation. Here again, the relations (2.288) and (2.289) would only be applied in the context of a computer program. For smaller problems, it is advisable to proceed from the elementary rotations and to obtain the Jacobian matrices $\boldsymbol{J}_{RiS}(t)$ and $\boldsymbol{J}'_{RiS}(t)$ appearing in (2.290) intuitively using a geometric approach.

When linearizing, we must especially take heed that $\boldsymbol{\eta} \cdot \delta\boldsymbol{\eta}$ is not a quadratic term in the sense of a Taylor series expansion. This means that the series expansion in (2.281) is required up to the second term in order to determine the virtual motion. If the series expansion in (2.281) is already interrupted after the first term, completely false results could emerge when determining the generalized forces. This connection is commonly overlooked in the literature and in the development of program systems designed to investigate linear multibody systems. If we obtain the linearized accelerations (2.286), (2.291) not by total derivation of the linearized velocities (2.285), (2.290), but rather using general, nonlinear relations (2.185), (2.200), then even the third term in (2.281) must be taken into account for $\dot{\boldsymbol{y}}_S(t) = \boldsymbol{0}$. This method is therefore not recommended for setting up linear relations.

Nonholonomic systems can be linearized without difficulty. In addition to (2.281), (2.220) must then also undergo a series expansion, i.e. the target motion is determined by $\boldsymbol{y}_S(t)$ and $\boldsymbol{z}_S(t)$.

Furthermore, it is often useful to carry out a partial linearization. In this case, some of the position coordinates and/or some velocity coordinates are viewed as small due to the physics or the actual motion, while the rest may be large. We then of course do not obtain completely linear equations of motion, but the solution can nonetheless be substantially simplified.

Example 2.17 (Overturning Pendulum). Let the target motion of the double pendulum, see Fig. 2.20, be given by the motion of the reference frame. Then the following applies with respect to the inertial frame

$$\boldsymbol{y}_S(t) = \begin{bmatrix} \Omega t \\ \Omega t \end{bmatrix}, \qquad \boldsymbol{\eta}(t) = \begin{bmatrix} \alpha_1 \\ \alpha_2 \end{bmatrix} \tag{2.292}$$

and it should be assumed here that, despite the large guidance motion $\boldsymbol{y}_S(t)$, only small deviations from this arise, i.e., $\alpha_1 \ll 1$, $\alpha_2 \ll 1$. The series expansion for the first location vector is written up to the second term

$${}_I\boldsymbol{r}_1 = \begin{bmatrix} 0 \\ \sin\Omega t \\ -\cos\Omega t \end{bmatrix} L + \begin{bmatrix} 0 \\ \alpha_1\cos\Omega t \\ \alpha_1\sin\Omega t \end{bmatrix} L + \begin{bmatrix} 0 \\ -\frac{1}{2}\alpha_1^2\sin\Omega t \\ \frac{1}{2}\alpha_1^2\cos\Omega t \end{bmatrix} L \tag{2.293}$$

and the Jacobian matrix is obtained for the first mass point in the form

$$ {}_I\boldsymbol{J}_{T1S} = \begin{bmatrix} 0 & 0 \\ \cos\Omega t & 0 \\ \sin\Omega t & 0 \end{bmatrix} L, \tag{2.294} $$

$$ {}_I\dot{\boldsymbol{J}}_{T1s} = \begin{bmatrix} 0 & 0 \\ -\sin\Omega t & 0 \\ \cos\Omega t & 0 \end{bmatrix} L, \tag{2.295} $$

$$ {}_I\boldsymbol{v}_{is} = \begin{bmatrix} 0 \\ \Omega\cos\Omega t \\ \Omega\sin\Omega t \end{bmatrix}. \tag{2.296} $$

In accordance with (2.285), (2.286), now the velocity and acceleration of the first mass point are also determined. We then obtain for example

$$ {}_I\boldsymbol{v}_1 = \begin{bmatrix} 0 \\ \dot\alpha_1\cos\Omega t - \alpha_1\Omega\sin\Omega t + \Omega\cos\Omega t \\ \dot\alpha_1\sin\Omega t + \alpha_1\Omega\cos\Omega t + \Omega\sin\Omega t \end{bmatrix} L. \tag{2.297} $$

If we observe the pendulum motion in the reference frame, then all expressions are simplified. Equations (2.293)–(2.297) yield with (2.270) Equations (2.298)–(2.299), *End of Example 2.17*,

$$ {}_R\boldsymbol{r}_1 = \begin{bmatrix} 0 \\ \alpha_1 \\ -1 + \frac{1}{2}\alpha_1^2 \end{bmatrix} L, \qquad {}_R\boldsymbol{J}_{T1S} = \begin{bmatrix} 0 & 0 \\ 1 & 0 \\ 0 & 0 \end{bmatrix} L \tag{2.298} $$

and

$$ {}_R\boldsymbol{v}_1 = \begin{bmatrix} 0 \\ \dot\alpha_1 + \Omega \\ \alpha_1\Omega \end{bmatrix} L. \tag{2.299} $$

Kinematics is a very extensive branch of applied dynamics. Many important concepts and definitions have been introduced in this chapter, such as the point, rigid body, and continuum models, the motion types of translation, rotation, and strain, generalized coordinates and velocities, holonomic and nonholonomic constraints, motion relative to reference frames, and small, linearizable deviations from a target motion. All these basic concepts will be made use of again and again in the following chapters.

Chapter 3
Basic Kinetics

The motion of mechanical systems is caused and influenced by forces and torques. Kinetics describes the effect of forces and torques on free systems. Constrained systems are thus assigned to free systems in accordance with the method of sections. Starting from the kinetics of the point, we will proceed to the kinetics of the rigid body and of a continuum. The basic equations of kinetics, the Newton and Euler equations, constitute, together with the law of conservation of mass, the mechanical balance equations. We also call Newton's equations the principle of conservation of linear momentum or the momentum balance and Euler's equations the principle of conservation of angular momentum, the balance of moment of momentum, or angular momentum balance. Momentum balance and the angular momentum balance are independent basic equations for a continuum, while for free point systems both equations are mutually convertible. The momentum balance and angular momentum balance can also be supplemented by the energy balance, which is however dependent on the first two principals in the case of isothermal processes in a continuum. Nevertheless, simpler solutions are at hand in many cases by using the energy balance.

3.1 Kinetics of a Mass Point

The motion of a mass point is changed by the forces acting on it. This is expressed specifically in Newton's second law. Furthermore, every force between two mass points causes a corresponding counterforce, which is stated in Newton's third law.

3.1.1 Newton Equations

Newton's second law is the foundation of mass point kinetics. Newton [39] formulated this law of motion in the year 1687 as follows: '*Mutationem motus*

W. Schiehlen and P. Eberhard, *Applied Dynamics*, DOI 10.1007/978-3-319-07335-4_3,

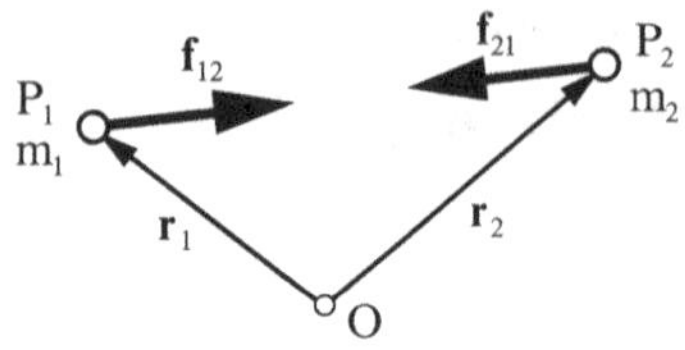

Fig. 3.1 System of two mass points

proportionalem esse vi motrici impressae, et fieri secundum lineam rectam qua vis illa imprimitur', which translates into 'The rate of change of momentum of a body is proportional to the resultant force acting on the body and happens in the same direction'. We can immediately see that Newton's formulation permits far-reaching interpretations, which can also be found numerously in the literature, see e.g. Szabo [63]. If we make use of calculus, Newton's second law can be written in the form

$$m\boldsymbol{a} = m\frac{d\boldsymbol{v}}{dt} = \boldsymbol{f}, \tag{3.1}$$

where m is the mass of a mass point, $\boldsymbol{a}$ the 3×1 vector of the absolute acceleration in an inertial frame, and $\boldsymbol{f}$ the 3×1 vector of all forces acting on the mass point. Newton's first law results from (3.1) as a special case. With $\boldsymbol{f} = \boldsymbol{0}$, we obtain $m\boldsymbol{v} =$ constant, which is also called the law of inertia.

Newton's third law, the law of reaction, is already concerned with a mass point system, see Fig. 3.1. If a force is applied between two mass points P_1 and P_2, then its action on both mass points is equal but opposed. Thus

$$\boldsymbol{f}_{12} = -\boldsymbol{f}_{21}. \tag{3.2}$$

If we now apply (3.1) to the two mass points shown in Fig. 3.1, we then obtain with (3.2)

$$m_1\boldsymbol{a}_1 + m_2\boldsymbol{a}_2 = \boldsymbol{0}. \tag{3.3}$$

This means that the force action on the total system disappears. This reveals the relationship between system demarcation and the type of acting forces. The law of reaction is also called the method of sections, since the internal forces of a system can be turned into external forces by changing the demarkation. For a system of p mass points, Newton's equations are written

$$m_i\boldsymbol{a}_i(t) = \boldsymbol{f}_i(t), \qquad i = 1(1)p, \tag{3.4}$$

where the explanations given for (3.1) are analogously valid.

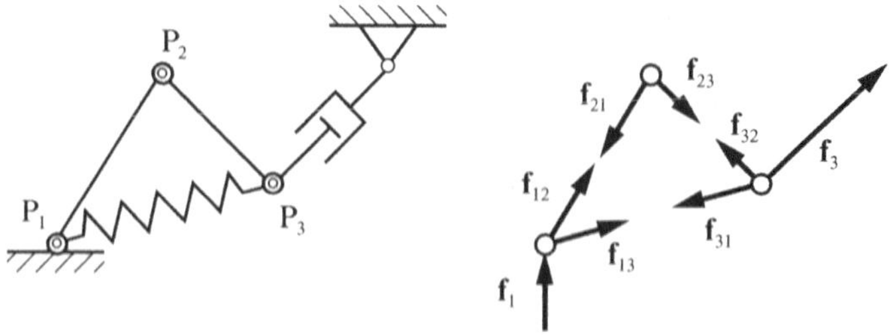

Fig. 3.2 Internal and external forces acting on a point system

3.1.2 Types of Forces

Newton's third law has made it clear that the forces in a system require a more precise classification. Figure 3.2 shows a constrained mass point system with $f = 3 \cdot 3 - 3 = 6$ degrees of freedom. The three mass points are connected to each other by two rods and a spring. Point P_1 is sliding on a frictionless surface, and point P_3 is suspended on a damper. This system contains the external forces $\boldsymbol{f}_1$ and $\boldsymbol{f}_3$ and the internal forces $\boldsymbol{f}_{12}, \boldsymbol{f}_{13}, \boldsymbol{f}_{21}, \boldsymbol{f}_{23}, \boldsymbol{f}_{31}, \boldsymbol{f}_{32}$. The applied forces $\boldsymbol{f}_{13}, \boldsymbol{f}_{31}$, and $\boldsymbol{f}_3$ derive from force laws and obey the spring characteristics and the damper characteristics, the reaction forces $\boldsymbol{f}_1, \boldsymbol{f}_{12}, \boldsymbol{f}_{21}, \boldsymbol{f}_{23}, \boldsymbol{f}_{32}$ originate from bearings and represent reactions to the motion restriction caused by the constraints. The external forces always appear once, internal forces always in pairs. However, all forces represent external forces from the standpoint of a single point.

In a mass point system, we therefore distinguish, according to the selected system boundaries, between external forces $\boldsymbol{f}_i^a$ and internal forces $\boldsymbol{f}_i^i$. Or we can make a distinction based on the origin of the forces between applied forces $\boldsymbol{f}_i^e$ and reaction forces $\boldsymbol{f}_i^r, i = 1(1)p$. Following the method of sections, in order to apply Newton's laws (3.4) all forces are converted into external forces acting on the point under consideration,

$$\boldsymbol{f}_i = \boldsymbol{f}_i^a + \boldsymbol{f}_i^i = \boldsymbol{f}_i^e + \boldsymbol{f}_i^r, \qquad i = 1(1)p. \tag{3.5}$$

The external forces $\boldsymbol{f}_i^a$ have their cause outside the selected system boundaries. The internal forces of a system

$$\boldsymbol{f}_i^i = \sum_{j=1}^{p} \boldsymbol{f}_{ij}, \qquad i = 1(1)p, \tag{3.6}$$

obey the reaction law

$$\boldsymbol{f}_{ij} + \boldsymbol{f}_{ji} = \mathbf{0}, \qquad i, j = 1(1)p. \tag{3.7}$$

Applied forces are derived from force elements such as springs and dampers and are known functions of time and of initially unknown motions and reactions of the system. Reaction forces are determined by constraints. They belong like motion quantities to unknowns of the system.

Reaction forces are reducible to generalized reaction forces by means of distribution matrices. The number of generalized reaction forces is equal to the number q of holonomic constraints in a system. So the $q \times 1$ vector of the generalized reaction forces can be introduced,

$$\boldsymbol{g} = \begin{bmatrix} g_1 \; g_2 \; \dots \; g_q \end{bmatrix} . \tag{3.8}$$

For these generalized reaction forces, we can then determine the $3 \times q$ distribution matrices $\boldsymbol{F}_i(\boldsymbol{y},t)$, and the following yields for reaction forces in holonomic systems

$$\boldsymbol{f}_i^r = \boldsymbol{F}_i(\boldsymbol{y},t) \cdot \boldsymbol{g}(t), \qquad i = 1(1)p. \tag{3.9}$$

As opposed to the $q < 3p$ generalized reaction forces $\boldsymbol{g}$, the $3p$ unknown reaction forces $\boldsymbol{f}_i^r$ are thus linearly dependent on each other. The distribution matrices $\boldsymbol{F}_i$ can be obtained either intuitively from geometrical considerations or analytically from the constraints of the system. The first possible method is based on an analysis in a Cartesian frame. It gives the generalized reaction forces a direct mechanical meaning. The second possibility will be discussed in detail in Chap. 4 in connection with the principles of mechanics.

Applied forces are categorized as ideal forces, which are not dependent on reaction forces, and non-ideal contact forces. Ideal forces are further subdivided into P-forces, PD-forces, and PI-forces. The proportional P-forces are dependent on position and time

$$\boldsymbol{f}_i^e = \boldsymbol{f}_i^e(\boldsymbol{x},t). \tag{3.10}$$

Besides the conservative spring forces, gravity forces, and the purely time-dependent control forces are also considered P-forces. The proportional-differential PD-forces depend also on velocity

$$\boldsymbol{f}_i^e = \boldsymbol{f}_i^e(\boldsymbol{x},\dot{\boldsymbol{x}},t). \tag{3.11}$$

A typical example of PD-forces is a spring strut, which corresponds to the parallel connection of a damper and a spring. The proportional-integral PI-forces are determined by the position and the position integrals of the system as well,

$$\boldsymbol{f}_i^e = \boldsymbol{f}_i^e(\boldsymbol{x},\boldsymbol{w},t), \qquad \dot{\boldsymbol{w}} = \dot{\boldsymbol{w}}(\boldsymbol{x},\boldsymbol{w},t). \tag{3.12}$$

The position integrals are formed with the $h \times 1$ vector $\boldsymbol{w}(t)$ of forces. Series connection of a damper and a spring leads to PI-forces, just like the eigen dynamics of an actuator.

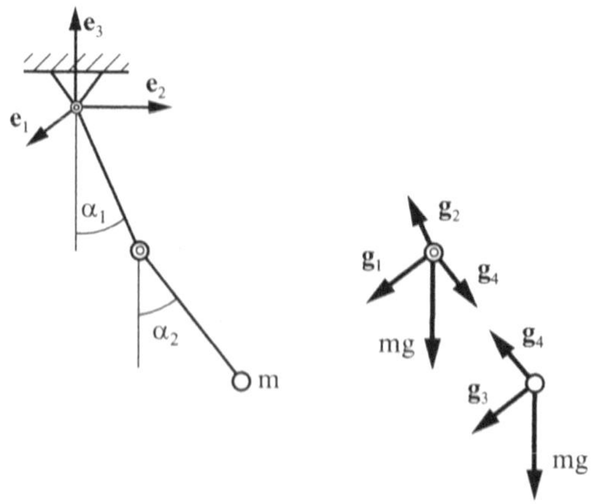

Fig. 3.3 Weights and reaction forces acting on a double pendulum

The ideal forces (3.10)–(3.12) occur in free and frictionless constrained systems. If there are also constraints involving friction, then non-ideal friction forces also come into play

$$\boldsymbol{f}_i^e = \boldsymbol{f}_i^e(\boldsymbol{y},\boldsymbol{g},t). \tag{3.13}$$

These friction forces are, just like contact forces, characterized by the coupling of applied forces and reaction forces. This coupling can lead to considerable complexity in the subsequent solution of the motion equation.

In engineering, individual types of forces are realized by corresponding construction elements, of which a few can be seen in Fig. 1.3.

Example 3.1 (Forces Acting on a Double Pendulum). A double pendulum, see also Example 2.9, is a spatial two-point system with four degrees of freedom. For the four constraints (2.181) there are four generalized reaction forces

$$\boldsymbol{g} = \begin{bmatrix} g_1 \; g_2 \; g_3 \; g_4 \end{bmatrix}. \tag{3.14}$$

Figure 3.3 shows the generalized reaction forces. The internal reaction force g_4 has been plotted twice to assist visualization. The reaction forces g_1, g_3 are parallel to the basis vector e_1. We thereby obtain, with (3.9), the distribution matrices

$$\boldsymbol{F}_1(\boldsymbol{y}) = \begin{bmatrix} 1 & 0 & 0 & 0 \\ 0 & -\sin\alpha_1 & 0 & \sin\alpha_2 \\ 0 & \cos\alpha_1 & 0 & -\cos\alpha_2 \end{bmatrix}, \tag{3.15}$$

$$\boldsymbol{F}_2(\boldsymbol{y}) = \begin{bmatrix} 0 & 0 & 1 & 0 \\ 0 & 0 & 0 & -\sin\alpha_2 \\ 0 & 0 & 0 & \cos\alpha_2 \end{bmatrix}. \tag{3.16}$$

The six coordinates of the reaction forces $\boldsymbol{f}_1^r$ and $\boldsymbol{f}_2^r$ are thus determined by four generalized reaction forces.

Also, the applied gravity forces are written

$$\boldsymbol{f}_1^e = \begin{bmatrix} 0 \\ 0 \\ -mg \end{bmatrix}, \quad \boldsymbol{f}_2^e = \begin{bmatrix} 0 \\ 0 \\ -mg \end{bmatrix}, \tag{3.17}$$

with which we have determined all forces acting on the double pendulum.

End of Example 3.1.

3.2 Kinetics of a Rigid Body

The motion of a rigid body is determined by the forces and torques acting upon it. In the case of a free rigid body, forces influence translation, and moments affect rotation related to the center of mass. Yet this kinematic decoupling is generally lost with constrained rigid bodies. In order to investigate the motion of a rigid body, we need Euler equations in addition to the Newton equations. This is analogously true for multibody systems.

A rigid body is a continuum, a model that was used by Euler with great success. In his 1775 paper '*Nova methodus motum corporum rigidorum determinandi*', or 'A New Method for Determining the Motion of Rigid Bodies', the integral forms of momentum balance and angular momentum balance were introduced, after he had already published in 1758 the gyroscopic equations that today bear his name. It is therefore historically inaccurate to designate the momentum balance of the rigid body as 'Newton equations' and the angular momentum balance as 'Euler equations'. This designation will nonetheless be preserved in this book in order to illustrate the close relationship between point systems and multibody systems.

3.2.1 Newton and Euler Equations

A rigid body may consist of a compact aggregate of mass points, but still Newton equations generally do not suffice for its description. Figure 3.4 shows two rigid bodies composed of four mass points and five or six rigid joint bars, respectively. In the plane, each body has three degrees of freedom. The left-hand body can

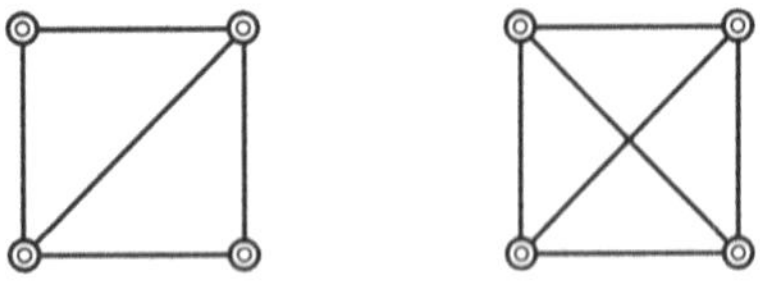

Fig. 3.4 Statically determinate and statically indeterminate point system

be calculated completely with the help of point mechanics. For three generalized coordinates and five generalized reaction forces, eight Newton equations are available. The right-hand body on the other hand cannot be described by means of point mechanics. The three generalized coordinates and the six generalized reaction forces cannot be calculated using the eight Newton equations. We also say that the left-hand body is statically determinate while the right-hand body is statically indeterminate. A rigid continuum is completely indeterminate statically, and its calculation thus always requires both Newton and Euler equations. The concept of static determinacy is derived from the field of stereo-statics. In the case of moving bodies, we can therefore also use the term kinetic determinacy. This usage is less common, however.

In continuum mechanics, the mass dm is assigned to the mass point P of the rigid body K, see Fig. 2.8. Thus, for the mass of the body we get

$$m = \int_K dm = \int_V \rho dV \tag{3.18}$$

where the integral should cover the volume V of the body K, and ρ denotes the density.

For the Newton equations of the rigid body, we obtain via integration over all mass points from (3.1) and (3.2) with (2.119)

$$\begin{aligned}&\int_K \boldsymbol{a}(t) + [\tilde{\boldsymbol{\alpha}}(t) + \tilde{\boldsymbol{\omega}}(t) \cdot \tilde{\boldsymbol{\omega}}(t)] \cdot \boldsymbol{r}_P(\boldsymbol{\rho}, t) dm \\ &\qquad = m\boldsymbol{a}(t) + [\tilde{\boldsymbol{\alpha}}(t) + \tilde{\boldsymbol{\omega}}(t) \cdot \tilde{\boldsymbol{\omega}}(t)] \cdot \int_K \boldsymbol{r}_P(\boldsymbol{\rho}, t) dm = \boldsymbol{f}(t),\end{aligned} \tag{3.19}$$

where $\boldsymbol{f}(t)$ represents the 3×1 vector of all external applied forces acting on the body K and $\boldsymbol{r}_P$ denotes the vector from the reference point P_1 to the mass points P. The position of the center of mass C with reference to P_1 is denoted by $\boldsymbol{r}_C$. If we select the center of mass C as the reference point P_1, then it yields

$$\boldsymbol{r}_C(t) = \frac{1}{m} \int_K \boldsymbol{r}_P(\boldsymbol{\rho}, t) dm \tag{3.20}$$

and with (2.79) we obtain

$$\boldsymbol{r}_C = \int_K \boldsymbol{r}_P(\rho, t) dm = \boldsymbol{S}(t) \cdot \int_K \boldsymbol{\rho} dm = \boldsymbol{0}. \tag{3.21}$$

Thus the Newton equations for the rigid body with respect to the center of mass are written in compliance with (3.19)

$$m\boldsymbol{a}(t) = \boldsymbol{f}(t). \tag{3.22}$$

These equations are also called the principle of linear momentum, since the acceleration of the center of mass is crucial. In the following, the center of mass – unless stated otherwise – will always be chosen as the reference point.

The Euler equations for an arbitrary body K read as

$$\int_K \tilde{\boldsymbol{r}} \cdot \ddot{\boldsymbol{r}} dm = \boldsymbol{l}, \tag{3.23}$$

where $\boldsymbol{l}$ is the 3×1 vector of all external moments acting on the body K. It should be noted in particular that, in the case of a body or continuum, the angular momentum $\int \tilde{\boldsymbol{r}} \cdot \dot{\boldsymbol{r}} dm$ and the total moment $\boldsymbol{l}$ are independent of the momentum $\int \dot{\boldsymbol{r}} dm$ and of the forces $\boldsymbol{f}$. The Newton equations and the Euler equations are fundamental, general, and mutually independent principles of mechanics. If we now apply (3.23) to the rigid body K, we then obtain with (2.80), (2.119), and (3.21)

$$\tilde{\boldsymbol{r}}(t) \cdot m\boldsymbol{a} + \int_K \tilde{\boldsymbol{r}}_P(\boldsymbol{\rho},t) \cdot [\tilde{\boldsymbol{\alpha}}(t) + \tilde{\boldsymbol{\omega}}(t) \cdot \tilde{\boldsymbol{\omega}}(t)] \cdot \boldsymbol{r}_P(\boldsymbol{\rho},t) dm = \boldsymbol{l}_O(t), \tag{3.24}$$

where $\boldsymbol{l}_O(t)$ represents the sum of all external moments with respect to the origin O of the frame. If we also insert (3.22) into (3.24), then we get

$$\int_K \tilde{\boldsymbol{r}}_P \cdot [\tilde{\boldsymbol{\alpha}} + \tilde{\boldsymbol{\omega}} \cdot \tilde{\boldsymbol{\omega}}] \cdot \boldsymbol{r}_P dm = \boldsymbol{l}_O - \tilde{\boldsymbol{r}} \cdot \boldsymbol{f} = \boldsymbol{l} \tag{3.25}$$

with the 3×1 vector $\boldsymbol{l}$ of the sum of all external moments in respect of the center of mass C.

To help convert the second term in the integral (3.25), we can use the relation

$$\tilde{\boldsymbol{x}} \cdot \tilde{\boldsymbol{x}} = \boldsymbol{x}\boldsymbol{x} - \boldsymbol{x} \cdot \boldsymbol{x}\boldsymbol{E}. \tag{3.26}$$

Application of (3.26) twice, with (A.18), yields the relation

$$\begin{aligned} \tilde{\boldsymbol{r}}_P \cdot \tilde{\boldsymbol{\omega}} \cdot \tilde{\boldsymbol{\omega}} \cdot \boldsymbol{r}_P &= \tilde{\boldsymbol{r}}_P \cdot (\boldsymbol{\omega}\boldsymbol{\omega} - \boldsymbol{\omega} \cdot \boldsymbol{\omega}\boldsymbol{E}) \cdot \boldsymbol{r}_P \\ &= \tilde{\boldsymbol{\omega}} \cdot (-\boldsymbol{r}_P\boldsymbol{r}_P + \boldsymbol{r}_P \cdot \boldsymbol{r}_P\boldsymbol{E}) \cdot \boldsymbol{\omega} = \tilde{\boldsymbol{\omega}} \cdot \tilde{\boldsymbol{r}}_P^T \cdot \tilde{\boldsymbol{r}}_P \cdot \boldsymbol{\omega}. \end{aligned} \tag{3.27}$$

From (3.25) we thus obtain

$$\int_K \tilde{\boldsymbol{r}}_P^T \cdot \tilde{\boldsymbol{r}}_P dm \cdot \boldsymbol{\alpha} + \tilde{\boldsymbol{\omega}} \cdot \int_K \tilde{\boldsymbol{r}}_P^T \cdot \tilde{\boldsymbol{r}}_P dm \cdot \boldsymbol{\omega} = \boldsymbol{l}. \tag{3.28}$$

If we introduce the 3×3 inertia tensor

$$\boldsymbol{I} = \int_K \tilde{\boldsymbol{r}}_P^T \cdot \tilde{\boldsymbol{r}}_P dm, \tag{3.29}$$

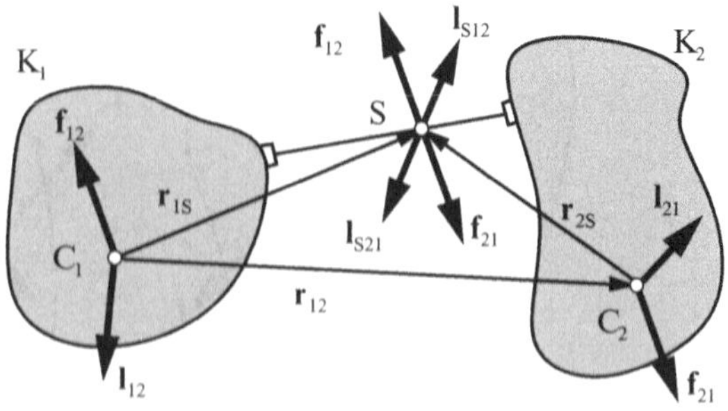

Fig. 3.5 System of two rigid bodies

then we finally arrive at the Euler equations

$$\boldsymbol{I}(t) \cdot \boldsymbol{\alpha}(t) + \tilde{\boldsymbol{\omega}}(t) \cdot \boldsymbol{I}(t) \cdot \boldsymbol{\omega}(t) = \boldsymbol{l}(t). \tag{3.30}$$

The Euler equations of the form (3.30) apply initially in the inertial frame, and the inertia tensor $\boldsymbol{I}(t)$ and external moments $\boldsymbol{l}(t)$ should be related to the center of mass C. Section 3.2.3 shows that the Euler equations (3.30) are also valid in a body-fixed frame.

Newton's third law must be extended by analogy to rigid bodies, see Fig. 3.5. The forces and torques acting between two bodies K_1 and K_2 of a multibody system are, at a common intersection S, equal and opposite

$$\boldsymbol{f}_{12} = -\boldsymbol{f}_{21}, \qquad \boldsymbol{l}_{S12} = -\boldsymbol{l}_{S21}. \tag{3.31}$$

The reaction law (3.31) is too unhandy for multibody systems because it is only valid for the common intersection S between two bodies. However, the Euler equations will have to be referred to the respective centers of mass, see (3.30). It is therefore more handy to write the reaction law with respect to the center of mass as well.

Then (3.31) becomes

$$\boldsymbol{f}_{12} = -\boldsymbol{f}_{21}, \qquad \boldsymbol{l}_{12} = -\boldsymbol{l}_{21} - \tilde{\boldsymbol{r}}_{12} \cdot \boldsymbol{f}_{21}, \tag{3.32}$$

where $\boldsymbol{r}_{12}$ is the location vector between the centers of mass C_1 and C_2 as shown in Fig. 3.5. If we now apply (3.22), (3.30) to a two-body system, then with (3.32) all the internal forces and torques drop out after assembling.

With a rigid body, we can also expose the internal forces and torques for a sectional plane by means of the method of sections, see Fig. 3.6. At each point Q of the sectional plane A, a 3×1 stress vector is acting

$$\boldsymbol{t} = \frac{d\boldsymbol{f}}{dA}, \tag{3.33}$$

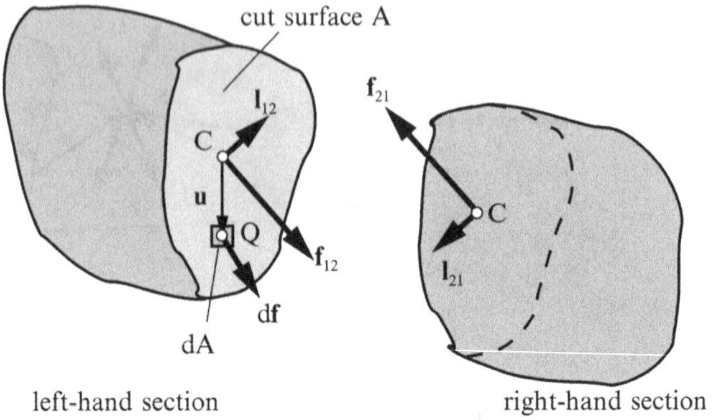

Fig. 3.6 Internal forces and torques of a rigid body

which refers to the surface element dA, whereby torque stresses are ignored in accordance with the definition of a nonpolar continuum, see Becker and Bürger [9]. With the 3×1 distance vector $\boldsymbol{u}$ lying in sectional area A, the cutting wrench is written

$$\boldsymbol{f}_{12} = \int_A \boldsymbol{t} dA, \qquad \boldsymbol{l}_{12} = \int_A \tilde{\boldsymbol{u}} \cdot \boldsymbol{t} dA. \tag{3.34}$$

The area centroid C is taken here as the reference point.

The cutting wrench $(\boldsymbol{f}_{12}, \boldsymbol{l}_{12})$ can generally be determined by means of the Newton and Euler equations for exposed body sections. On the other hand, determining the stress vector $\boldsymbol{t}$, and thus the stress tensor as well, from the cutting wrench alone is not possible because a wrench cannot be uniquely resolved into a system of forces. In order to calculate the distribution of stress in a body, we must depart from the rigid body model and replace it by an elastic continuum. This clarifies the limits of the model of a rigid body. It is suitable for determining large motions that apply both locally for an infinitesimal element and globally for the entire body, while the stress distribution inside the body remains indeterminate. But even given a rigid body, we can assume a linear distribution of stress in the sectional area of the cutting wrench, as is customary, e.g. in engineering beam bending. Further particulars will be discussed in Sect. 5.4.2 in connection with estimating strength in a multibody system. On the other hand, all properties of a continuum apply that are derived from the balance equations, to rigid bodies also. For example, the symmetry of the stress tensor results from the Euler equations if the latter are applied to an infinitesimal rigid tetrahedral element.

For a multibody system of p rigid bodies, the Newton-Euler equations are written as

$$m_i \boldsymbol{a}_i(t) = \boldsymbol{f}_i(t), \tag{3.35}$$

$$\boldsymbol{I}_i(t) \cdot \boldsymbol{\alpha}_i(t) + \tilde{\boldsymbol{\omega}}_i(t) \cdot \boldsymbol{I}_i(t) \cdot \boldsymbol{\omega}_i(t) = \boldsymbol{l}_i(t), \qquad i = 1(1)p. \tag{3.36}$$

The forces and torques acting on each particular body can be subdivided into external and internal forces and torques of the multibody system or, using another classification, into applied and reaction forces,

$$\boldsymbol{f}_i = \boldsymbol{f}_i^a + \boldsymbol{f}_i^i = \boldsymbol{f}_i^e + \boldsymbol{f}_i^r, \tag{3.37}$$

$$\boldsymbol{l}_i = \boldsymbol{l}_i^a + \boldsymbol{l}_i^i = \boldsymbol{l}_i^e + \boldsymbol{l}_i^r. \tag{3.38}$$

External forces $\boldsymbol{f}_i^a$ and external torques $\boldsymbol{l}_i^a$ have their causes outside of the boundaries of the multibody system under consideration. Internal forces and torques

$$\boldsymbol{f}_i^i = \sum_{j=1}^{p} \boldsymbol{f}_{ij}, \qquad \boldsymbol{l}_i^i = \sum_{j=1}^{p} \boldsymbol{l}_{ij} \tag{3.39}$$

obey the reaction law (3.7) of forces and the reaction law

$$\boldsymbol{l}_{ij} + \boldsymbol{l}_{ji} + \tilde{\boldsymbol{r}}_{ij} \cdot \boldsymbol{f}_{ji} = \boldsymbol{0}, \qquad i,j = 1(1)p, \tag{3.40}$$

for the torques related to the center of mass C_i. Also, the location vector $\boldsymbol{r}_{ij}$ between the centers of mass C_i and C_j appears in (3.40) with $i,j = 1(1)p$.

For the applied forces $\boldsymbol{f}_i^e$ and applied torques $\boldsymbol{l}^e$, all relations (3.10)–(3.13) derived for forces in point systems apply accordingly.

Furthermore, with the $(q+r) \times 1$ vector $\boldsymbol{g}(t)$ of the generalized reaction forces, the following applies for reactions in nonholonomic systems,

$$\boldsymbol{f}_i^r = \boldsymbol{F}_i(\mathbf{y},\mathbf{z},t) \cdot \boldsymbol{g}(t), \qquad \boldsymbol{l}_i^r = \boldsymbol{L}_i(\mathbf{y},\mathbf{z},t) \cdot \boldsymbol{g}(t), \tag{3.41}$$

i.e. the $3 \times (q+r)$ distribution matrices $\boldsymbol{F}_i$ and $\boldsymbol{L}_i$ are, in nonholonomic systems, dependent not only on the $f \times 1$ position vector $\mathbf{y}(t)$, but they can also be functions of the $g \times 1$ velocity vector $\mathbf{z}(t)$. On the other hand, this additional dependence is dropped for holonomic systems, see e.g. (3.9).

3.2.2 *Mass Geometry of a Rigid Body*

The mass geometry of a rigid body is described by the inertia tensor (3.29). The inertia tensor is a time-invariant parameter in the reference configuration or in the body-fixed frame $\{C;\boldsymbol{e}_\alpha\}$, respectively,

$$\boldsymbol{I} = \boldsymbol{S}^T \cdot \int_K \tilde{\boldsymbol{r}}_P^T \cdot \tilde{\boldsymbol{r}}_P dm \cdot \boldsymbol{S} = \int_K \tilde{\boldsymbol{\rho}}^T \cdot \tilde{\boldsymbol{\rho}} dm = \text{constant}. \tag{3.42}$$

With (3.26), we can also write it as

$$\boldsymbol{I} = \int_K (\boldsymbol{\rho} \cdot \boldsymbol{\rho} \boldsymbol{E} - \boldsymbol{\rho}\boldsymbol{\rho}) dm \tag{3.43}$$

or in coordinates

$$\boldsymbol{I} = \begin{bmatrix} I_{11} & I_{12} & I_{31} \\ I_{12} & I_{22} & I_{23} \\ I_{31} & I_{23} & I_{33} \end{bmatrix}. \tag{3.44}$$

The inertia tensor is a symmetric, generally positive definite tensor. Its diagonal elements $I_{\alpha\alpha}$, $\alpha = 1(1)3$ are called moments of inertia, and its remaining elements I_{12}, I_{23}, I_{31} are referred to as moments of deviation. Important inequalities are valid for the moments of inertia

$$I_{11} + I_{22} > I_{33}, \qquad I_{22} + I_{33} > I_{11}, \qquad I_{33} + I_{11} > I_{22}, \tag{3.45}$$

which are also called triangle inequalities. Analogous inequalities are also valid for the sides of a triangle. In degenerated cases, the inequalities become equalities. For rod-shaped bodies, the inertia tensor then becomes positively semidefinite.

Like any positive definite tensor, the inertia tensor also has three positive eigenvalues, the principal moments of inertia I_α, $\alpha = 1(1)3$. The principal moments of inertia are obtained from the eigenvalue problem

$$(\lambda \boldsymbol{E} - \boldsymbol{I}) \cdot \boldsymbol{x} = \boldsymbol{0} \tag{3.46}$$

where λ is an initially unknown eigenvalue and $\boldsymbol{x}$ the eigenvector belonging to λ. The solution of the eigenvalue problem is based on the characteristic equation

$$\det(\lambda \boldsymbol{E} - \boldsymbol{I}) = \lambda^3 - a_1 \lambda^2 + a_2 \lambda - a_3 = 0 \tag{3.47}$$

with the base invariants

$$a_1 = \mathrm{Sp}\boldsymbol{I} = I_{11} + I_{22} + I_{33} = I_1 + I_2 + I_3, \tag{3.48}$$

$$a_2 = I_{11}I_{22} + I_{22}I_{33} + I_{33}I_{11} - I_{12}^2 - I_{23}^2 - I_{31}^2 = I_1 I_2 + I_2 I_3 + I_3 I_1, \tag{3.49}$$

$$a_3 = \det \boldsymbol{I}, \tag{3.50}$$

which always have the same values in any frame, and therefore in the principal axis frame as well. The roots of (3.47) thus correspond to the principal moments of inertia. If we insert the principal moments of inertia I_i sequentially into (3.46), we obtain the three eigenvectors $\boldsymbol{x}_i$. These eigenvectors are perpendicular to each other or can be selected orthogonal to each other in the case of multiple eigenvalues. The eigenvectors thus define a Cartesian frame, the principal axis frame $\{C; \boldsymbol{e}_{H\alpha}\}$. The time-invariant transformation matrix between the body-fixed frame K and the principal axis frame H can be set up using the normalized eigenvectors

$$\boldsymbol{S}_{HK} = \left[\boldsymbol{x}_1 \; \boldsymbol{x}_2 \; \boldsymbol{x}_3 \right], \qquad \boldsymbol{x}_\alpha \cdot \boldsymbol{x}_\alpha = 1. \tag{3.51}$$

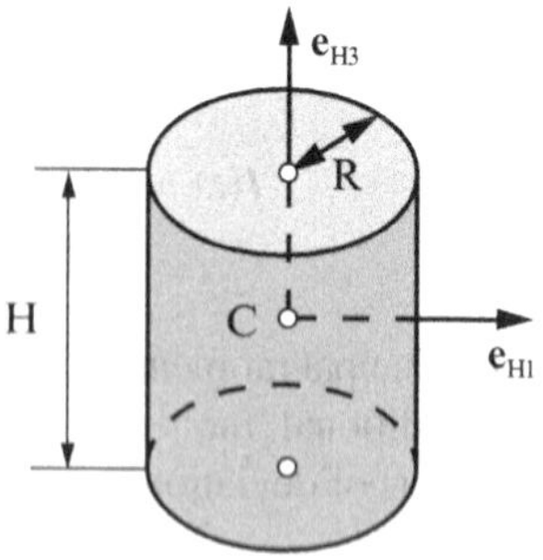

Fig. 3.7 Circular cylinder in a body-fixed frame

In the principal axis frame, the inertia tensor has a diagonal shape

$$_H\boldsymbol{I} = \boldsymbol{diag}\{I_1 \;\; I_2 \;\; I_3\}. \tag{3.52}$$

We thus use if possible the principal axis frame as the body-fixed frame. Proceeding from this, we then find the time-dependent inertia tensor in the inertial frame

$$\boldsymbol{I}(t) = \boldsymbol{S}(t) \cdot \boldsymbol{S}_{KH} \cdot \boldsymbol{diag}\{I_1 \;\; I_2 \;\; I_3\} \cdot \boldsymbol{S}_{KH}^T \cdot \boldsymbol{S}^T(t). \tag{3.53}$$

With the help of the principal moments of inertia, the mass-geometric properties of a rigid body can even be represented in one image plane only. To this end, we need the dimensionless inertia parameters

$$k_1 = \frac{I_2 - I_3}{I_1}, \qquad k_2 = \frac{I_3 - I_1}{I_2}, \qquad k_3 = \frac{I_1 - I_2}{I_3}, \tag{3.54}$$

which enable us to plot a characteristic point in the Magnus shape triangle, see Magnus [35].

Example 3.2 (Inertia Tensor in the Inertial Frame). For the homogeneous circular cylinder, see Fig. 3.7, of radius R and height H, we obtain the principal axis frame $H, \{C, \boldsymbol{e}_H\}$ the principal moments of inertia

$$I_1 = I_2 = m\frac{3R^2 + H^2}{12}, \qquad I_3 = m\frac{R^2}{2}. \tag{3.55}$$

It should rotate with small deviations $\alpha(t) \ll 1, \beta(t) \ll 1$ around its 3-axis with angular speed Ω. Then the linearized rotation tensor is written

$$\boldsymbol{S}(t) = \begin{bmatrix} \cos\Omega t & -\sin\Omega t & \beta \\ \sin\Omega t & \cos\Omega t & -\alpha \\ \alpha\sin\Omega t & \alpha\cos\Omega t & \\ -\beta\cos\Omega t & +\beta\sin\Omega t & 1 \end{bmatrix}. \tag{3.56}$$

From (3.53) with $\boldsymbol{S}_{KH} = \boldsymbol{E}$, we obtain for the inertia tensor in the inertial frame

$$\boldsymbol{I}(t) = \begin{bmatrix} I_1 & 0 & (I_3 - I_1)\beta \\ 0 & I_1 & -(I_3 - I_1)\alpha \\ (I_3 - I_1)\beta & -(I_3 - I_1)\alpha & I_3 \end{bmatrix}. \tag{3.57}$$

The principal moments of inertia remain constant as a result of the small deviations, but additional moments of position do occur, and as a result time-dependent moments of deviations as well.

End of Example 3.2.

3.2.3 Relative Motion of the Frame

The Newton-Euler equations (3.35) are valid in the inertial frame. Yet with the kinematic relations of relative motion, see Sect. 2.4.1, these equations can also be given in a moving reference frame. With (2.257), we first obtain

$$m_i(\boldsymbol{r}_R^{**} + (\dot{\tilde{\boldsymbol{\omega}}}_R + \tilde{\boldsymbol{\omega}}_R \cdot \tilde{\boldsymbol{\omega}}_R) \cdot \boldsymbol{r}_{Ri} + 2\tilde{\boldsymbol{\omega}}_R \cdot \dot{\boldsymbol{r}}_{Ri} + \ddot{\boldsymbol{r}}_{Ri}) = \boldsymbol{f}_i \tag{3.58}$$

for the Newton equations, where the 3×1 location vector $\boldsymbol{r}_{Ri}$ points to the center of mass C_i of the body $K_i, i = 1(1)p$. After longer calculation, from (2.256), (2.258) we obtain the following result for the Euler equations,

$$\boldsymbol{I}_i \cdot \dot{\boldsymbol{\omega}}_R + \tilde{\boldsymbol{\omega}}_R \cdot \boldsymbol{I}_i \cdot \boldsymbol{\omega}_R + \tilde{\boldsymbol{\omega}}_R \cdot \boldsymbol{\omega}_{Ri} Sp \boldsymbol{I}_i + 2\tilde{\boldsymbol{\omega}}_{Ri} \cdot \boldsymbol{I}_i \cdot \boldsymbol{\omega}_R + \boldsymbol{I}_i \cdot \dot{\boldsymbol{\omega}}_{Ri} + \tilde{\boldsymbol{\omega}}_{Ri} \cdot \boldsymbol{I}_i \cdot \boldsymbol{\omega}_{Ri} = \boldsymbol{l}_i. \tag{3.59}$$

However, this equation can also be derived directly from the basic equation (3.23). The first two terms in (3.59) denote the moments of guidance acceleration, the middle two reveal the influence of Coriolis acceleration, and the last two describe the moments of relative rotational acceleration. In particular, we can see in (3.59) that the simple form (3.30) of the angular momentum balance is valid not only in the inertial frame but also in the body-fixed reference frame ($\boldsymbol{\omega}_{Ri} = \boldsymbol{0}$). For this reason, the body-fixed principal axis frame is always selected as the reference frame when investigating gyroscopic problems. Then (3.59) yields the dynamic Euler equations

$$\begin{aligned} \dot{\omega}_1(t) - k_1\omega_2(t)\omega_3(t) &= l_1(t)/I_1, \\ \dot{\omega}_2(t) - k_2\omega_3(t)\omega_1(t) &= l_2(t)/I_2, \\ \dot{\omega}_3(t) - k_3\omega_1(t)\omega_2(t) &= l_3(t)/I_3 \end{aligned} \tag{3.60}$$

with dimensionless, time-invariant inertia parameters (3.54). In the case of multi-body systems however, the body-fixed principal axis frame becomes meaningless as a reference frame, since then only one inertia tensor is time-invariant, while all the others are again functions of time. The only option then would be to select, as shown in Sect. 2.4.1, a special reference frame R_j for each body K_j. This technique is made use of in the context of recursive formalisms, see Sect. 5.7.2. Yet this again increases the complexity involved in setting up equations of motion. In the final analysis, the inertial frame usually proves to be the simplest reference frame for multibody systems.

3.3 Kinetics of a Continuum

In contrast to a rigid body, the motion of a continuum is determined not only by concentrated forces and torques but also by continuously distributed force fields. In the context of applied dynamics, we waive the introduction of torque fields, i.e. only nonpolar continua are considered, the mass points of which cannot carry out rotations. The Newton-Euler equations then yield the Cauchy equations of motion, which, together with a material law, allow the investigation of motion. Even with the simplest material law, Hooke's law for linear-elastic material, closed solutions of the partial differential equations are only rarely found. For this reason, approximation methods, especially the method of finite elements, are very important.

3.3.1 Cauchy Equations

A few extensions must first be made for the nonrigid body K, see Fig. 3.8. While in the case of the rigid body only the resulting concentrated forces and torques are important, continua require that we take into account the continuously distributed force fields

$$\boldsymbol{f}(t) = \int_V \rho \boldsymbol{f}(\boldsymbol{\rho}, t) dV + \int_A \boldsymbol{t}(\boldsymbol{\rho}, t) dA + \Sigma \boldsymbol{f}_j(t). \tag{3.61}$$

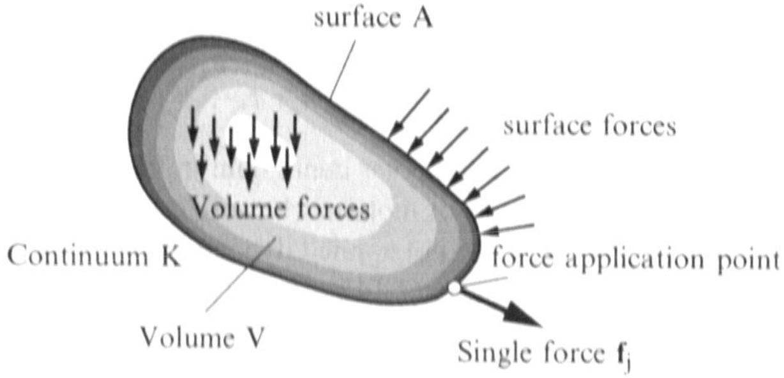

Fig. 3.8 Surface and volume forces acting on a continuum

In addition to the concentrated forces $\boldsymbol{f}_j(t)$, volume forces are introduced with the 3×1 vector of inertia force density $\boldsymbol{f}$ and the surface forces with the 3×1 stress vector $\boldsymbol{t}$. The integrals extend over the volume V and the surface A of the continuum K. As a rule, volume forces are external applied forces such as the weight. Surface forces are external applied forces if they are based on physical laws, e.g. on wind loading. However, external reaction forces can also come into play if the continuum is mounted on its surface. Internal surface forces, which are exposed with the method of sections, are, in free continua, also applied forces depending on the material law. The stresses $\boldsymbol{t}$ depend, via the 3×3 stress tensor $\boldsymbol{T}(\boldsymbol{\rho},t)$, on the cutting direction, which is characterized by the 3×1 normal vector $\boldsymbol{n}$,

$$\boldsymbol{t}(\boldsymbol{\rho},t) = \boldsymbol{n} \cdot \boldsymbol{T}(\boldsymbol{\rho},t), \tag{3.62}$$

see Becker and Bürger [9] or Lai, Rubin, Krempl [31].

If we now write the Newton equations (3.1) for a mass point of mass $dm = \rho dV$ and integrate over the entire body, we obtain, observing (3.61) and (3.62) with the Gauss law of vector analysis, the momentum balance

$$\int_V \rho \boldsymbol{a} dV = \int_V \rho \boldsymbol{f} dV + \int_V \mathrm{div} \boldsymbol{T} dV + \Sigma \boldsymbol{f}_j. \tag{3.63}$$

This equation is now valid not only for a finite continuum but also for an infinitesimal body, i.e. a mass point of the continuum,

$$\rho \boldsymbol{a} = \rho \boldsymbol{f} + \mathrm{div} \boldsymbol{T}. \tag{3.64}$$

If we also apply the Euler equations, we obtain as a sole additional predication the symmetry of the stress tensor

$$\boldsymbol{T} = \begin{bmatrix} \sigma_{11} & \tau_{12} & \tau_{31} \\ \tau_{12} & \sigma_{22} & \tau_{23} \\ \tau_{31} & \tau_{23} & \sigma_{33} \end{bmatrix} = \boldsymbol{T}^T, \tag{3.65}$$

where σ_{ii} represents the normal stresses and τ_{ij} the shear stresses. Six important elements thus remain again, which can be merged into a 6×1 stress vector

$$\boldsymbol{\sigma} = \begin{bmatrix} \sigma_{11} & \sigma_{22} & \sigma_{33} & \tau_{12} & \tau_{23} & \tau_{31} \end{bmatrix}. \tag{3.66}$$

The Newton-Euler equations for a nonpolar continuum in the form (3.64), (3.65) are also known as Cauchy equations of motion. They are essential basic equations of continuum mechanics in the context of applied dynamics.

With the differential operator matrix (2.142) introduced in kinematics, the basic equations (3.64) and (3.65) can be summarized compactly,

$$\rho \boldsymbol{a} = \rho \boldsymbol{f} + \mathscr{V}^T \cdot \boldsymbol{\sigma}, \tag{3.67}$$

where the symmetry of the stress tensor has also been taken into account.

3.3.2 Hooke's Material Law

The Cauchy equations of motion (3.64) and (3.65), which with (2.153) and (2.159) represent differential equations for the current deformation $\boldsymbol{r}(\boldsymbol{\rho},t)$, cannot be solved, since the stress tensor $\boldsymbol{T}(\boldsymbol{\rho},t)$ is initially still unknown. The stresses have to be expressed as a function of deformation with the material law.

The most important material law for applied dynamics is the linear Hooke's material law. It corresponds in 1D to a proportional applied spring force and represents generally a linear relation between the stresses and the strains. Instead of the tensorial formulation of Hooke's material law, which associates tensors of the second order for stresses and rotations with the tensor of the fourth order of the material, here a correspondingly re-sorted matrix representation will be selected. The following applies

$$\boldsymbol{\sigma} = \boldsymbol{H} \cdot \boldsymbol{e} \tag{3.68}$$

with the symmetric 6×6 matrix $\boldsymbol{H}$ of Hooke's law,

$$\boldsymbol{H} = \left[\begin{array}{ccc|ccc} 1-\nu & \nu & \nu & & & \\ \nu & 1-\nu & \nu & & \mathbf{0} & \\ \nu & \nu & 1-\nu & & & \\ \hline & & & \frac{1-2\nu}{2} & 0 & 0 \\ & \mathbf{0} & & 0 & \frac{1-2\nu}{2} & 0 \\ & & & 0 & 0 & \frac{1-2\nu}{2} \end{array}\right] \frac{E}{(1+\nu)(1-2\nu)}, \tag{3.69}$$

where E is Young's modulus and ν Poisson ratio.

Example 3.3 (Tension Rod). In a tension rod (cross section A) with an axial load (force F), a one-dimensional state of stress is given

$$\boldsymbol{\sigma} = \begin{bmatrix} \sigma \; 0 \; 0 \; 0 \; 0 \; 0 \end{bmatrix} \tag{3.70}$$

with the normal stress $\sigma = F/A$. The corresponding strain vector is

$$\boldsymbol{e} = \begin{bmatrix} 1 \; -\nu \; -\nu \; 0 \; 0 \; 0 \end{bmatrix} \varepsilon, \tag{3.71}$$

where lateral strain is taken into account. By inserting the material (3.70), (3.71) into (3.68), (3.69), we confirm the one-dimensional Hooke law

$$\sigma = E\varepsilon \tag{3.72}$$

in its simplest form.

End of Example 3.3.

3.3.3 Reaction Stress

In addition to applied stresses, constraint stresses can also arise in the context of a continuum. We differentiate here between external and internal reactions.

External reaction stresses are based on the external constraints (2.211). They are calculated by considering the explicit constraints on the surface of the continuum

$$\boldsymbol{r} = \boldsymbol{r}(\boldsymbol{\rho}, t) \qquad \text{on } A^r, \tag{3.73}$$

where A^r is the constrained part of the surface. By solving the Cauchy equations of motion we then obtain the initially unknown 3×1 stress vector $\boldsymbol{t}^r$ on A^r.

Internal reaction stresses are generated by internal constraints, see (2.209). The 3×3 reaction stress tensor has the form

$$\boldsymbol{T}^r = \frac{\partial \phi}{\partial \boldsymbol{F}} \cdot \boldsymbol{F}^T g(\boldsymbol{\rho}, t), \tag{3.74}$$

where $g(\boldsymbol{\rho}, t)$ is a generalized constraint stress distribution. Inasmuch as it can be calculated at all, the generalized constraint stress distribution is determined by the Cauchy equations of motion. The derivation of the relation (3.74) can be found e.g. in Becker and Bürger [9].

Chapter 4
Principles of Mechanics

The basic kinetic equations for points, bodies, and continua are immediately valid for free systems. These basic equations allow the calculation of motions when the forces and torques are given, or the resulting forces and torques can be determined from the motions. We can thus on the one hand calculate Kepler's first law from the Newton equations and the force of gravity, while on the other hand Newton's law of gravity can be obtained from the motion of the planets.

In constrained systems, initially unknown reaction forces and torques arise in addition to applied forces and torques. These reaction forces and torques can be determined in statically or kinetically defined systems with the help of the basic kinetic equations, but they do not directly influence motion, which can only take place in non-restricted directions. It therefore seems reasonable to eliminate the reaction forces and torques from the basic kinetic equations. This can be achieved with the help of the principles of mechanics.

Proceeding from the principle of virtual work, we will first discuss the principles of d'Alembert, Jourdain, and Gauss. The principle of minimal potential energy and Hamilton's principle will then be introduced. This will be followed by a derivation of the Lagrange equations of the first and second kind from the d'Alembert principle.

4.1 Principle of Virtual Work

In constrained systems, reaction forces and torques occur. As a result of the constraints, these forces and torques do not perform any virtual work.

The following applies for the virtual work of reaction forces acting on a mass point

$$\delta W^r = \boldsymbol{f}^r \cdot \delta \boldsymbol{r} = 0. \tag{4.1}$$

W. Schiehlen and P. Eberhard, *Applied Dynamics*, DOI 10.1007/978-3-319-07335-4_4,

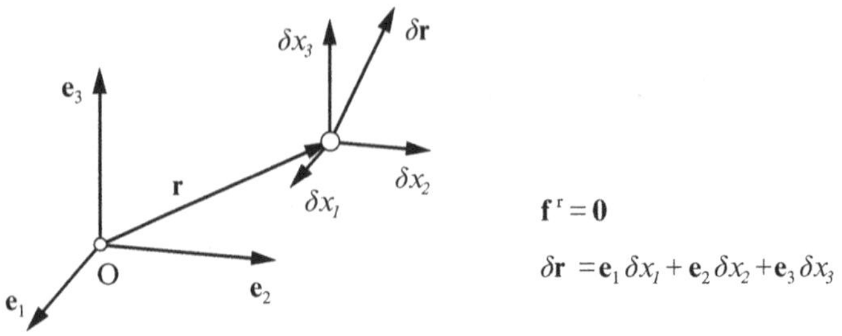

Fig. 4.1 Calculation of the virtual work of a free point

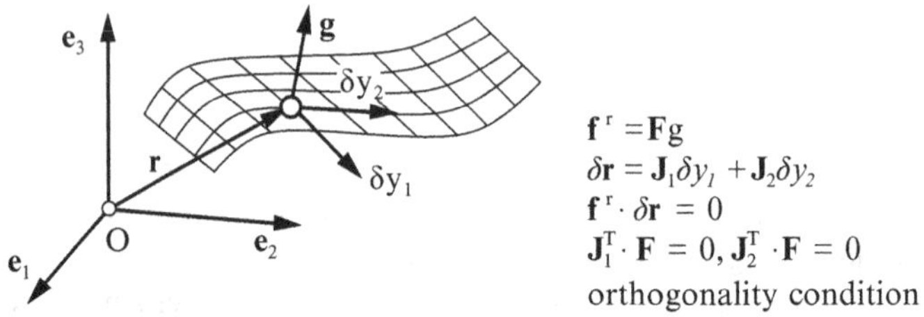

Fig. 4.2 Calculation of the virtual work of a point constrained in one normal direction

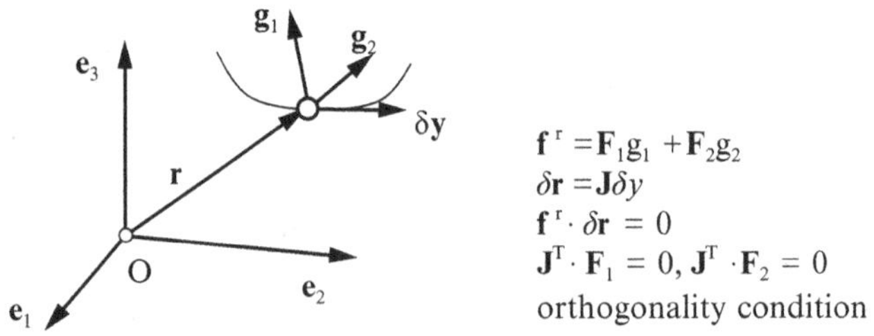

Fig. 4.3 Calculation of the virtual work of a point constrained in two normal directions

Here, δW^r is the virtual work of the reaction forces, $\boldsymbol{f}^r$ designates the 3×1 vector of the reaction forces, and $\delta\boldsymbol{r}$ is the 3×1 vector of the virtual displacements, which denote infinitesimal motions compatible with the constraints. The relation (4.1) is quite generally valid for free and arbitrarily constrained points, as shown by the following examples. In the case of a free point, see Fig. 4.1, the virtual work vanishes, since there is no reaction force. If the point is bound to a surface, see Fig. 4.2, or a curve, see Fig. 4.3, the virtual work vanishes as a result of the orthogonality of virtual motion and reaction force (Orthogonality condition). A point which is fixed in a statically definite manner, see Fig. 4.4, cannot carry out motion. For this reason, virtual work is again equal to zero. Thus, (4.1) represents the basis for the formulation of general mechanical principles.

In a system of p mass points, the virtual work of the reaction forces must likewise vanish,

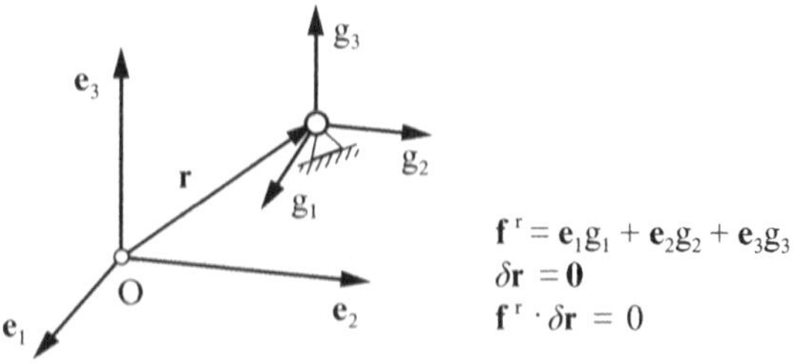

Fig. 4.4 Calculation of the virtual work of a statically fixed point

$$\delta W^r = \sum_{i=1}^{p} \boldsymbol{f}_i^r \cdot \delta \boldsymbol{r}_i = 0. \tag{4.2}$$

Also, we can now distinguish between external and internal reaction forces. Thus, according to (3.5) and (3.6)

$$\boldsymbol{f}_i^r = \boldsymbol{f}_i^{ra} + \sum_{j=1}^{p} \boldsymbol{f}_{ij}^r, \tag{4.3}$$

where we must still take into account the reaction law (3.7). If we insert (4.3) in (4.2), we then obtain with (3.7)

$$\delta W^r = \sum_{i=1}^{p} \boldsymbol{f}_i^{ra} \cdot \delta \boldsymbol{r}_i + \sum_{i=1}^{p} \sum_{j=i}^{p} \boldsymbol{f}_{ij}^r \cdot (\delta \boldsymbol{r}_i - \delta \boldsymbol{r}_j) = 0. \tag{4.4}$$

We can see that the virtual displacements $\delta \boldsymbol{r}_i$ are decisive for the virtual work of the external reaction forces, while the virtual work of the internal reaction forces is determined by the relative virtual displacement differences $(\delta \boldsymbol{r}_i - \delta \boldsymbol{r}_j)$.

In addition, we find for the virtual work on a nonpolar continuum

$$\delta W^r = \int_K d\boldsymbol{f}^r \cdot \delta \boldsymbol{r} = 0, \tag{4.5}$$

where $d\boldsymbol{f}^r$ are the reaction forces acting on a mass point P of mass dm. If we now consider that the inertia force density $\boldsymbol{f}$ is based on applied forces, then (3.64) results in the relation

$$\delta W^r = \int_V (\mathrm{div} \boldsymbol{T}^r) \cdot \delta \boldsymbol{r} dV = 0. \tag{4.6}$$

This volume integral can now be converted with the product rule of vector analysis and Gauss's law, applied respectively to $\mathrm{div}(\boldsymbol{T}^r \cdot \delta \boldsymbol{r})$. We then obtain for the virtual work

$$\delta W^r = \int_A \boldsymbol{t}^r \cdot \delta \boldsymbol{r} dA - \int_V \mathrm{Sp}(\boldsymbol{T}^r \cdot \delta \boldsymbol{G}) dV = \int_A \boldsymbol{t}^r \cdot \delta \boldsymbol{r} dA - \int_V \boldsymbol{\sigma}^{rT} \cdot \delta \boldsymbol{e} dV = 0, \tag{4.7}$$

where (2.141) and (3.66) have been used for the sake of simplicity. In addition to the virtual displacements $\delta\boldsymbol{r}$ on the surface A, in (4.7) the virtual strains $\delta\boldsymbol{G}$ or $\delta\boldsymbol{e}$, respectively, arise in the volume V of the continuum K. The respective first term in (4.7) corresponds to the virtual work of external reaction forces, while the respective second term describes the virtual work of internal reaction forces. This allocation was already found in point systems, see (4.4). Thus, in a continuum without internal constraints, only the first term remains, and the integral can only be calculated using that section of the surface whose freedom of motion is restricted by constraints or connections.

The virtual work of the reaction forces and torques is now also provided for a polar continuum,

$$\delta W^r = \int_K (d\boldsymbol{f}^r \cdot \delta\boldsymbol{r} + d\boldsymbol{l}^r \cdot \delta\boldsymbol{s}) = 0. \tag{4.8}$$

Here, $d\boldsymbol{f}$ and $d\boldsymbol{l}$ are the reaction forces and torques acting on the volume element of the mass point P of mass dm. The virtual displacement $\delta\boldsymbol{r}$ must be supplemented by the virtual rotation $\delta\boldsymbol{s}$ in the case of polar continua. Multibody systems, due to the individual torques, belong to the nonpolar continua, too. If we consider the kinematics of a rigid body, (4.8) results in

$$\delta W^r = \sum_{i=1}^{p}\left[\boldsymbol{f}_i^r \cdot \delta\boldsymbol{r}_i + \int_{K_i}(d\boldsymbol{f}_i^r \cdot \delta\tilde{\boldsymbol{s}}_i \cdot \boldsymbol{r}_p + d\boldsymbol{l}_i^r \cdot \delta\boldsymbol{s}_i)\right] = \sum_{i=1}^{p}(\boldsymbol{f}_i^r \cdot \delta\boldsymbol{r}_i + \boldsymbol{l}_i^r \cdot \delta\boldsymbol{s}_i) = 0 \tag{4.9}$$

with the virtual motion $\delta\boldsymbol{r}_i, \delta\boldsymbol{s}_i$ of the multibody system, which is given by (2.201). The effects of the forces $d\boldsymbol{f}_i^r$ with the lever arms $\boldsymbol{r}_P(\boldsymbol{\rho})$ and the torques $d\boldsymbol{l}_i^r$ are contained in the resulting reaction torques $\boldsymbol{l}_i^r$.

In the case of constrained systems, the vanishing virtual work of the reaction forces corresponds to a general orthogonality relation. We will work this out for all aforementioned systems using point systems as a representative example.

The virtual motions of point systems have already been discussed in detail in the context of kinematics, see (2.188). The normal directions $\boldsymbol{n}_{ik}$ decisive for the reaction forces $\boldsymbol{f}_i^r$ are obtained either from geometrical intuition or by calculation from the implicit form (2.175) of the constraints. With the 3×1 vector

$$\boldsymbol{n}_{ik} = \frac{\partial\phi_k}{\partial\boldsymbol{x}} \cdot \frac{\partial\boldsymbol{x}}{\partial\boldsymbol{r}_i}, \qquad k = 1(1)q, \tag{4.10}$$

and the generalized reaction force g_k belonging to the constraint ϕ_k, we obtain for the resulting reaction force $\boldsymbol{f}_i^r$ acting on the mass point P_i the sum

$$\boldsymbol{f}_i^r = \sum_{k=1}^{q}\boldsymbol{f}_{ik}^r = \sum_{k=1}^{q}\frac{\partial\phi_k}{\partial\boldsymbol{x}} g_k \cdot \frac{\partial\boldsymbol{x}}{\partial\boldsymbol{r}_i}. \tag{4.11}$$

The normal vector (4.10) is not normalized. By division with $|\boldsymbol{n}_{ik}|$, we can obtain, if required, the direction of the normal unit vector. By normalizing the normal vector, the generalized reaction force g_k becomes equal in amount to the corresponding reaction force $\boldsymbol{f}^r_{ik}$. Insertion into (4.2) yields

$$\sum_{i=1}^{p}\left[\left(\sum_{k=1}^{q} g_k \frac{\partial \phi_k}{\partial \boldsymbol{x}}\right) \cdot \frac{\partial \boldsymbol{x}}{\partial \boldsymbol{r}_i} \cdot \frac{\partial \boldsymbol{r}_i}{\partial \boldsymbol{x}} \cdot \sum_{\ell=1}^{f} \frac{\partial \boldsymbol{x}}{\partial y_\ell} \delta y_\ell\right] = 0. \tag{4.12}$$

If we now introduce the $f \times 1$ position vector $\boldsymbol{y}$ from (2.177), the virtual change in position $\delta\boldsymbol{y}$ and the Jacobian matrices $\boldsymbol{H}_{Ti}$, $\boldsymbol{I}$, and $\boldsymbol{J}_{Ti}$ according to (2.188) and (2.189) as well as the $q \times 1$ vector of the generalized reaction forces (3.8), and if we define the functional matrices

$$\boldsymbol{F}_i^T = \frac{\partial \boldsymbol{\phi}}{\partial \boldsymbol{r}_i}, \qquad \boldsymbol{G}^T = \frac{\partial \boldsymbol{\phi}}{\partial \boldsymbol{x}}, \qquad \boldsymbol{H}_{Ti}^{+} = \frac{\partial \boldsymbol{x}}{\partial \boldsymbol{r}_i}, \tag{4.13}$$

then we can also write (4.12) as

$$\boldsymbol{g} \cdot \left(\sum_{i=1}^{p} \boldsymbol{F}_i^T \cdot \boldsymbol{J}_{Ti}\right) \cdot \delta\boldsymbol{y} = 0. \tag{4.14}$$

If we finally merge the functional matrices into the global $3p \times q$ distribution matrix $\overline{\boldsymbol{Q}}$ and the global $3p \times f$ Jacobian matrix $\overline{\boldsymbol{J}}$ of the point system,

$$\overline{\boldsymbol{Q}}^T = \left[\boldsymbol{F}_1^T \, \boldsymbol{F}_2^T \cdots \boldsymbol{F}_p^T\right], \qquad \overline{\boldsymbol{J}} = \begin{bmatrix} \boldsymbol{J}_{T1} \\ \boldsymbol{J}_{T2} \\ \vdots \\ \boldsymbol{J}_{Tp} \end{bmatrix}, \tag{4.15}$$

then (4.14) becomes the orthogonality relation

$$\overline{\boldsymbol{Q}}^T \cdot \overline{\boldsymbol{J}} = \boldsymbol{0}, \qquad \overline{\boldsymbol{J}}^T \cdot \overline{\boldsymbol{Q}} = \boldsymbol{0}. \tag{4.16}$$

Due to $\Sigma \boldsymbol{H}_{Ti}^{+} \cdot \boldsymbol{H}_{Ti} = p\boldsymbol{E}$ it yields also

$$\boldsymbol{G}^T \cdot \boldsymbol{I} = \boldsymbol{0}, \qquad \boldsymbol{I}^T \cdot \boldsymbol{G} = \boldsymbol{0}. \tag{4.17}$$

The orthogonality relation can be written both in Cartesian coordinates (4.16) and in generalized coordinates (4.17). The reaction forces thus do not carry out – independently of our choice of coordinates – any virtual work, what applies to all mechanical systems.

The principle of virtual work, often call the principle of virtual displacement, can now be easily derived for static point systems. The equilibrium conditions of statics

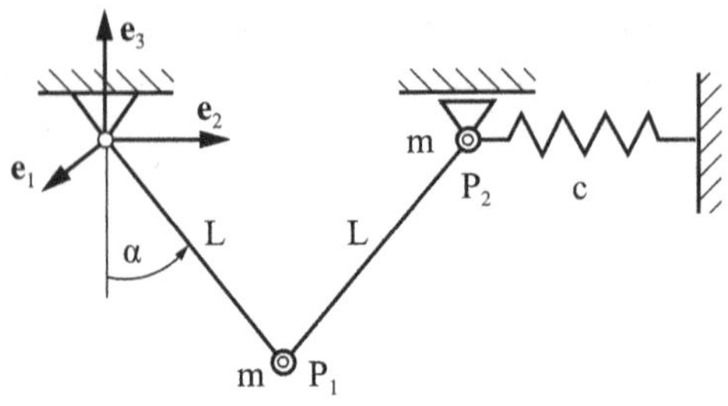

Fig. 4.5 Mechanism with spring support

demand that the sum of all external forces acting on each individual mass point vanishes

$$\boldsymbol{f}_i^a = \boldsymbol{0}, \qquad i = 1(1)p. \tag{4.18}$$

All external forces can be subdivided into applied external and reaction forces as a result of constraints,

$$\boldsymbol{f}_i^{ae} + \boldsymbol{f}_i^{ar} = \boldsymbol{0}, \qquad i = 1(1)p. \tag{4.19}$$

We thus obtain from (4.2) for the virtual work of a point system

$$\delta W^e = \sum_{i=1}^{p} \boldsymbol{f}_i^{ae} \cdot \delta \boldsymbol{r}_i = 0. \tag{4.20}$$

The principle of virtual work (4.20) thus states: A point system is then and only then in static equilibrium if the virtual work of the external applied forces vanishes. The great advantage for engineering applications is that equilibrium can be investigated without calculating the reaction forces.

Example 4.1 (Mechanism). The mechanism in Fig. 4.5 is a system of two mass points (mass m) with $f = 2 \cdot 3 - 5 = 1$ degrees of freedom. Let the spring (spring constant c) be unstrained in the horizontal position of the rods, $\alpha = 90°$. The position of equilibrium can now be easily determined with the principle of virtual work. The virtual displacements are

$$\delta \boldsymbol{r}_1 = \begin{bmatrix} 0 \\ L\cos\alpha \\ L\sin\alpha \end{bmatrix} \delta\alpha, \qquad \delta \boldsymbol{r}_2 = \begin{bmatrix} 0 \\ 2L\cos\alpha \\ 0 \end{bmatrix} \delta\alpha. \tag{4.21}$$

The weights (gravity g) and the spring force are the only applied forces

$$\boldsymbol{f}_1^e = \begin{bmatrix} 0 \\ 0 \\ -mg \end{bmatrix}, \qquad \boldsymbol{f}_2^e = \begin{bmatrix} 0 \\ 2cL(1-\sin\alpha) \\ -mg \end{bmatrix}. \tag{4.22}$$

The principle of virtual work directly provides the equilibrium condition

$$\delta W^e = (-mgL\sin\alpha + 4cL^2(1-\sin\alpha)\cos\alpha)\delta\alpha = 0 \tag{4.23}$$

or $mgL\sin\alpha + 4cL^2(1-\sin\alpha)\cos\alpha = 0$. For $mg = 4cL$, we obtain the numerical values $\alpha_1 = 27.97^\circ$ and $\alpha_2 = -117.97^\circ$.

The principle of virtual work applies not only to point system but also quite generally to all static mechanical systems. The relation (4.20) for the calculation of the work of applied forces must then be extended, however. The procedure necessary for this will be illustrated in the next section using the example of the d'Alembert principle, which is equivalent to an expansion of the principle of virtual work to dynamic mechanical systems.

End of Example 4.1.

4.2 Principles of d'Alembert, Jourdain, and Gauss

In this section, the principles will first be specified for point systems and then extended to multibody systems and continua. From the Newton equations (3.4), with (3.5) in compliance with (4.2), we obtain the d'Alembert principle in the Lagrange form

$$\sum_{i=1}^{p}(m_i\boldsymbol{a}_i - \boldsymbol{f}_i^e)\cdot\delta\boldsymbol{r}_i = 0. \tag{4.24}$$

Remarkable, although often overlooked, is the fact that only the applied forces appear in the d'Alembert principle, not all external forces. The d'Alembert principle thus allows – in compliance with the principle of virtual work – the establishment of equations of motions without direct consideration of the reaction forces. Yet there are systems, in which the applied forces depend on the reaction forces, e.g. in the case of friction forces. Then the reaction forces have an indirect influence on the motion, which may indeed affect the solution, but not the setting up of equations of motion.

The d'Alembert principle is valid for all holonomic systems. The virtual motions (2.188) appearing in (4.24) underline this. Moreover, one can show that the d'Alembert principle is also valid for linear nonholonomic systems. But it is easier and straightforward to work with the Jourdain principle in the case of nonholonomic systems,

$$\sum_{i=1}^{p}(m_i\boldsymbol{a}_i - \boldsymbol{f}_i^e)\cdot\delta'\boldsymbol{v}_i = 0. \tag{4.25}$$

The Jourdain principle (4.25) states that the virtual power of the reaction forces vanishes

$$\delta P^r = \sum_{i=1}^{p} \boldsymbol{f}_i^r \cdot \delta' \boldsymbol{v}_i = 0. \tag{4.26}$$

The Jourdain principle is thus closely related with the d'Alembert principle (4.24). Instead of virtual motions, the Jourdain principle deals with virtual velocity variations (2.231). The implicit holonomic constraints (2.175) are supplemented by the implicit nonholonomic constraints (2.219), which in compliance with (4.10) can be used to calculate the nonholonomic reaction forces.

In addition, we can also introduce virtual accelerations

$$\delta'' \boldsymbol{r} = \delta'' \boldsymbol{v} = \boldsymbol{0}, \qquad \delta'' t = 0, \qquad \delta'' \boldsymbol{a} \neq \boldsymbol{0}. \tag{4.27}$$

We can then write the Gauss principle

$$\sum_{i=1}^{p} (m_i \boldsymbol{a}_i - \boldsymbol{f}_i^e) \cdot \delta'' \boldsymbol{a}_i = 0. \tag{4.28}$$

An intuitive explanation of the Gauss principle is that the least constraint defined by the averaged acceleration deviations is minimized. The Gauss principle has so far not acquired much further engineering significance.

The d'Alembert principle for multibody systems follows from the Newton-Euler equations (3.35) under consideration of (3.37) and (4.9) in the form

$$\sum_{i=1}^{p} [(m_i \boldsymbol{a}_i - \boldsymbol{f}_i^e) \cdot \delta \boldsymbol{r}_i + (\boldsymbol{I}_i \cdot \boldsymbol{\alpha}_i + \tilde{\boldsymbol{\omega}}_i \cdot \boldsymbol{I}_i \cdot \boldsymbol{\omega}_i - \boldsymbol{l}_i^e) \cdot \delta \boldsymbol{s}_i] = 0. \tag{4.29}$$

In addition to the virtual displacements $\delta \boldsymbol{r}_i$, multibody systems require that we take into account the virtual rotations $\delta \boldsymbol{s}_i$ as well, which together represent the virtual motion according to (2.201). The Jourdain principle (4.25) for multibody systems can be written in a completely analogous manner. Then the virtual velocity changes and virtual rotation velocity changes (2.232) come into effect. This is especially advantageous in the case of nonholonomic systems.

In a continuum, inertia and volume forces are applied forces, and stresses can be categorized as applied stresses $\boldsymbol{T}^e$ and reaction stresses $\boldsymbol{T}^r$. From the Cauchy equations of motion (3.64), after multiplication with the virtual displacement and integration over the volume V of the body K at hand, we obtain the d'Alembert principle

$$\int_V (\rho \boldsymbol{a} - \rho \boldsymbol{f} - \mathrm{div} \boldsymbol{T}^e) \cdot \delta \boldsymbol{r} dV = 0, \tag{4.30}$$

where the reaction stresses are dropped in conformity with (4.6). If we carry out a conversion corresponding to (4.7), we arrive at the d'Alembert principle in the form

$$\int_V [(\rho \boldsymbol{a} - \rho \boldsymbol{f}) \cdot \delta \boldsymbol{r} + \boldsymbol{\sigma}^e \cdot \delta \boldsymbol{e}] dV - \int_A \boldsymbol{t}^e \cdot \delta \boldsymbol{r} dA = 0. \tag{4.31}$$

A final third form is reminiscent of the principle of virtual work

$$\int_V \rho \boldsymbol{a} \cdot \delta \boldsymbol{r} - \delta W^e = 0, \tag{4.32}$$

where the virtual work of the applied forces appearing in (4.30) and (4.31) is summarized in the term δW^e.

There is a difference between the representations (4.30) and (4.31) that is essential for dynamics. This becomes clear when the stress vector $\boldsymbol{\sigma}$ is expressed by Hooke's law (3.68) and the strain vector $\boldsymbol{e}$ from (2.143) by the displacement vector $\boldsymbol{w}$

$$\int_V (\rho \boldsymbol{a} - \rho \boldsymbol{f} - \mathscr{V}^T \cdot \boldsymbol{H} \cdot \mathscr{V} \cdot \boldsymbol{w}) \cdot \delta \boldsymbol{r} dV = 0, \tag{4.33}$$

$$\int_V (\rho \boldsymbol{a} - \rho \boldsymbol{f}) \cdot \delta \boldsymbol{r} + (\mathscr{V} \cdot \boldsymbol{w}) \cdot \boldsymbol{H} \cdot \delta(\mathscr{V} \cdot \boldsymbol{w}) dV - \int_A (\boldsymbol{N}^T \cdot \boldsymbol{H} \cdot \mathscr{V} \cdot \boldsymbol{w}) \cdot \delta \boldsymbol{r} dA = 0. \tag{4.34}$$

In (4.33), the differential operator matrix $\mathscr{V}$ of strain is applied twice to the distribution vector $\boldsymbol{w}$, respectively, in (4.34) on the other hand only once. The consequence of this is that the solution for the location vector $\boldsymbol{r}$ or displacement vector $\boldsymbol{w}$ in (4.33) must satisfy the geometric and dynamic boundary conditions, while in (4.34) only the often simpler geometrical boundary conditions must be observed. The dynamic boundary conditions based on the forces are contained in (4.34) in the surface integral and are thus automatically satisfied. Equation (4.34) is therefore always employed for approximations, as, e.g., in the finite element method. On the other hand, (4.33) is used in the case of continuous systems.

For a clearer presentation, in (4.34) the 6×3 matrix

$$\boldsymbol{N} = \begin{bmatrix} n_1 & 0 & 0 & n_2 & 0 & n_3 \\ 0 & n_2 & 0 & n_1 & n_3 & 0 \\ 0 & 0 & n_3 & 0 & n_2 & n_1 \end{bmatrix} \tag{4.35}$$

was introduced to the 3×1 normal vector of the surface

$$\boldsymbol{n} = \begin{bmatrix} n_1 & n_2 & n_3 \end{bmatrix}. \tag{4.36}$$

The structure of (4.35) corresponds exactly to that of (2.142).

4.3 Principle of Minimum Potential Energy

In conservative systems, the applied forces are denoted by potentials. The total potential energy U of a mechanical system is given by the potential U_a of the external forces and the potential U_i of the internal forces

$$U = U_a + U_i. \tag{4.37}$$

In the case of a conservative continuum, which e.g. obeys the linear-elastic Hooke material law, the potential of the internal forces corresponds to the deformation energy

$$U_i = \frac{1}{2}\int_V \boldsymbol{\sigma} \cdot \boldsymbol{e} dV. \tag{4.38}$$

If we now vary the potential energy with respect to the virtual displacements, we then obtain with (4.20)

$$\delta U = \sum_{i=1}^{p} \frac{\partial U}{\partial \boldsymbol{r}_i} \cdot \delta \boldsymbol{r}_i = -\sum_{i=1}^{p} \boldsymbol{f}_i^e \cdot \delta \boldsymbol{r}_i = -\delta W^e = 0. \tag{4.39}$$

A conservative mechanical system is thus in a position of equilibrium if its total potential in this position is stationary. It can also be shown that this position of equilibrium is stable precisely if

$$U \overset{!}{=} \min \quad \rightarrow \quad \delta U = 0,\ \delta^2 U > 0 \quad \rightarrow \quad \delta W^e = 0,\ \delta^2 W_e < 0 \tag{4.40}$$

is valid. If we now assume a linear-elastic mechanical system, e.g. a Hooke body, then the total potential is a positive definite quadratic form. Then only one stable position of equilibrium with minimum potential energy exists. We have thus identified an important application area of the principle of minimum potential energy: conservative, linear-elastic, static systems.

The principle of minimum potential energy yields no benefits compared to the principle of virtual work. On the contrary, the quadratic form of a linear-elastic potential must be converted by differentiation into linear spring forces – an unnecessarily complex operation.

Example 4.2 (Mechanism). The applied external forces acting on the device in Fig. 4.5 (gravity g, spring constant c) have the potential

$$U = mgr_{13} + mgr_{23} + \frac{1}{2}c(2L - r_{22})^2. \tag{4.41}$$

The location vectors of the mass points are

$$r_1 = \begin{bmatrix} 0 \\ L\sin\alpha \\ -L\cos\alpha \end{bmatrix}, \qquad r_2 = \begin{bmatrix} 0 \\ 2L\sin\alpha \\ 0 \end{bmatrix}. \tag{4.42}$$

Inserted into (4.41), we have

$$U = -mgL\cos\alpha + 2cL^2(1-\sin\alpha)^2. \tag{4.43}$$

The first variation δU with respect to the generalized coordinates α yields the negative virtual work, see (4.23). Since in this case we do not have a linear-elastic mechanical system, we still need to check the stability of the equilibrium positions with (4.40). The second variation $\delta^2 U$ supplies $\delta^2 U(\alpha_1) > 0$, $\delta^2 U(\alpha_2) < 0$, i.e. there is one stable and one unstable position of equilibrium.

End of Example 4.2.

4.4 Hamilton Principle

The Hamilton principle represents the extension of the principle of minimum potential energy to conservative, dynamic systems. However, we generally do not assume linear-elastic material behavior in this context.

For a conservative system, (4.32) can also be written

$$\int_V \rho a \cdot \delta r dV + \delta U = 0. \tag{4.44}$$

If we now integrate (4.44) with the fixed boundaries t_0 and t_1, we first obtain

$$\int_{t_0}^{t_1}\int_V \rho a \cdot \delta r dV dt + \int_{t_0}^{t_1} \delta U dt = 0. \tag{4.45}$$

On the other hand, the following applies for the variation of the kinetic energy

$$\delta T = \int_V \rho \dot{r} \cdot \delta \dot{r} dV = \int_V \frac{\partial}{\partial t}(\rho \dot{r} \cdot \delta r) dV - \int_V \rho a \cdot \delta r dV, \tag{4.46}$$

and integration leads to

$$\int_{t_0}^{t_1} \delta T = \left. \int_V (\rho \dot{r} \cdot \delta r) dV \right|_{t_0}^{t_1} - \int_{t_0}^{t_1}\int_V \rho a \cdot \delta r dV dt. \tag{4.47}$$

If, in addition to (2.186) from the virtual displacements, we also require

$$\delta \boldsymbol{r}(t_0) = \boldsymbol{0}, \qquad \delta \boldsymbol{r}(t_1) = \boldsymbol{0}, \tag{4.48}$$

we then obtain from (4.45) and (4.47)

$$\delta \int_{t_0}^{t_1} (T - U) dt = \delta \int_{t_0}^{t_1} L dt = 0. \tag{4.49}$$

We have thereby found Hamilton's principle, an extremum principle, whereby

$$L = T - U \tag{4.50}$$

is the known Lagrange function. Hamilton's principle states that the magnitude $\int L dt$ designated as action function has a stationary value

$$\frac{d}{dt} \int_{t_0}^{t_1} L dt = 0. \tag{4.51}$$

This insight may have nature-philosophical significance, but in applied dynamics, (4.51) leads to the same results as the d'Alembert principle (4.32).

4.5 Lagrange Equations of the First Kind

In order to derive the Lagrange equations of the first kind for a holonomic point system, we can first write the d'Alembert principle (4.24) with (4.2) in the form

$$\sum_{i=1}^{p} (m_i \boldsymbol{a}_i - \boldsymbol{f}_i^e - \boldsymbol{f}_i^r) \cdot \delta \boldsymbol{r}_i = 0. \tag{4.52}$$

Since the virtual displacements $\delta \boldsymbol{r}_i$ are dependent on each other as a result of the q constraints, only $f = 3p - q$ variations $\delta \boldsymbol{r}_i$ can be freely selected, and the associated bracket terms must vanish. The remaining bracket terms vanish by means of a suitable choice of generalized reaction forces, which act here as Lagrange multipliers. We thus obtain the Lagrange equations of the first kind in the form

$$m_i \boldsymbol{a}_i(\boldsymbol{x}, \dot{\boldsymbol{x}}, \ddot{\boldsymbol{x}}) = \boldsymbol{f}_i^e(\boldsymbol{x}, \dot{\boldsymbol{x}}, t) + \boldsymbol{F}_i(\boldsymbol{x}, t) \cdot \boldsymbol{g}(t), \qquad i = 1(1)p, \tag{4.53}$$

if we introduce the $3p \times 1$ position vector $\boldsymbol{x}(t)$ and the $g \times 1$ vector $\boldsymbol{g}(t)$ of the generalized reaction forces. We see that the $3p$ equations (4.53) do not suffice in order to determine the $3p + q$ unknowns. Therefore, we still have to supplement (4.53) with the q algebraic equations $\boldsymbol{\phi}(\boldsymbol{x}, t) = \boldsymbol{0}$ from (2.175).

Thus the Lagrange equations of the first kind represent a strongly linked nonlinear system of algebraic equations and differential equations, which also has a still higher order. This means that the Lagrange equations of the first kind are difficult to solve numerically.

The number of equations can be formally reduced after differentiating (2.175)

$$\boldsymbol{\phi}(\boldsymbol{x},t) = \mathbf{0}, \tag{4.54}$$

$$\frac{\partial\boldsymbol{\phi}}{\partial\boldsymbol{x}}\cdot\dot{\boldsymbol{x}} + \frac{d\boldsymbol{\phi}}{dt} = \mathbf{0}, \tag{4.55}$$

$$\frac{\partial\boldsymbol{\phi}}{\partial\boldsymbol{x}}\cdot\ddot{\boldsymbol{x}} + \frac{d}{dt}\frac{\partial\boldsymbol{\phi}}{\partial\boldsymbol{x}}\cdot\dot{\boldsymbol{x}} + \frac{d^2\boldsymbol{\phi}}{dt^2} = \mathbf{0}. \tag{4.56}$$

If we solve (4.53) according to $\ddot{\boldsymbol{x}}(t)$, which is always possible, and insert into (4.56), the result is a conditional equation for the generalized reaction forces in the form

$$\boldsymbol{g} = \boldsymbol{g}(\boldsymbol{x},\dot{\boldsymbol{x}},t). \tag{4.57}$$

Then, from the coupled equations (4.53), (4.54) remains only the system

$$m_i\boldsymbol{a}_i(\boldsymbol{x},\dot{\boldsymbol{x}},\ddot{\boldsymbol{x}}) = \boldsymbol{f}_i^e(\boldsymbol{x},\dot{\boldsymbol{x}},t) + \boldsymbol{F}_i(\boldsymbol{x},t)\cdot\boldsymbol{g}(\boldsymbol{x},\dot{\boldsymbol{x}},t) \tag{4.58}$$

for the $3p$ unknowns $\boldsymbol{x}(t)$. Yet since the position vector $\boldsymbol{x}(t)$ still has mutually dependent coordinates due to the constraints, the starting conditions $\boldsymbol{x}(t_0), \dot{\boldsymbol{x}}(t_0)$ must be calculated from (4.54) and (4.55). Moreover, it should be noted that the system (4.58), as a result of the differentiations (4.55) and (4.56), has two zero eigenvalues and is thus singular. These singularities can indeed be removed using a method by Baumgarte, see e.g. Wittenburg [65], but then systematic errors may arise in the numerical integration. Regardless, according to (4.58) we must still solve $q = 3p - f$ extra differential equations. The Lagrange equations of the first kind are therefore often less recommendable.

4.6 Lagrange Equations of the Second Kind

Lagrange equations of the second kind of a holonomic conservative point system can also be obtained from the d'Alembert principle. Following an intermediate calculation, such as the one provided by Magnus and Müller-Slany [36], we find the relations

$$\sum_{i=1}^{p} m_i\boldsymbol{a}_i\cdot\delta\boldsymbol{r}_i = \left(\frac{d}{dt}\frac{\partial T}{\partial\dot{\boldsymbol{y}}} - \frac{\partial T}{\partial\boldsymbol{y}}\right)\cdot\delta\boldsymbol{y}, \tag{4.59}$$

$$\sum_{i=1}^{p} \boldsymbol{f}_i^e\cdot\delta\boldsymbol{r}_i = -\frac{\partial U}{\partial\boldsymbol{y}}\cdot\delta\boldsymbol{y}, \tag{4.60}$$

from which, with (4.24) and (4.50), directly follows the Lagrange equations of the second kind due to the independence of the virtual motion δy of the generalized coordinates

$$\frac{d}{dt}\frac{\partial L}{\partial \dot{y}} - \frac{\partial L}{\partial y} = \mathbf{0}. \tag{4.61}$$

However, (4.61) can also be obtained from (4.51), since the Lagrange equations are nothing else than the Euler-Lagrange equation of the variation problem (4.51). The Lagrange equations are thus valid not only for point systems but for all kinds of holonomic mechanical systems.

Despite their widespread use, Lagrange equations of the second kind are often too laborious for the generation of equations of motion. This can be seen already by looking at the example of a scleronomic point system. In this case, the kinetic energy is

$$\begin{aligned} T &= \frac{1}{2}\sum_{i=1}^{p} \mathbf{v}_i \cdot m_i \mathbf{v}_i = \frac{1}{2}\sum_{i=1}^{p} \dot{\mathbf{y}}(t)\cdot \boldsymbol{J}_{Ti}^T(\mathbf{y}) m_i \cdot \boldsymbol{J}_{Ti}(\mathbf{y})\cdot \dot{\mathbf{y}}(t) \\ &= \frac{1}{2}\dot{\mathbf{y}}(t)\cdot \overline{\boldsymbol{J}}^T(\mathbf{y})\cdot \overline{\overline{\boldsymbol{M}}}\cdot \overline{\boldsymbol{J}}(\mathbf{y})\cdot \dot{\mathbf{y}}(t), \end{aligned} \tag{4.62}$$

where the velocities from (2.184) and the global Jacobian and mass matrices according to (5.21) and (5.20) are taken into account. This yields for the partial derivatives of the kinetic energy

$$\frac{\partial T}{\partial \dot{y}} = \overline{\boldsymbol{J}}^T \cdot \overline{\overline{\boldsymbol{M}}} \cdot \overline{\boldsymbol{J}} \cdot \dot{\mathbf{y}}, \qquad \frac{\partial T}{\partial y} = \frac{\partial(\overline{\boldsymbol{J}}\cdot \dot{\mathbf{y}})}{\partial y} \cdot \overline{\overline{\boldsymbol{M}}} \cdot \overline{\boldsymbol{J}} \cdot \dot{\mathbf{y}}. \tag{4.63}$$

So first the factor $\frac{1}{2}$ drops out, since (4.62) is a quadratic form with respect to $\dot{\mathbf{y}}(t)$ and $\overline{\boldsymbol{J}}(\mathbf{y})$. The total derivative of (4.63) according to time results in

$$\frac{d}{dt}\frac{\partial T}{\partial \dot{y}} = \overline{\boldsymbol{J}}^T \cdot \overline{\overline{\boldsymbol{M}}} \cdot \overline{\boldsymbol{J}} \cdot \ddot{\mathbf{y}} + \overline{\boldsymbol{J}}^T \cdot \overline{\overline{\boldsymbol{M}}} \cdot \frac{\partial(\overline{\overline{\boldsymbol{M}}} \cdot \overline{\boldsymbol{J}} \cdot \dot{\mathbf{y}})}{\partial y} \cdot \dot{\mathbf{y}} + \frac{\partial(\overline{\boldsymbol{J}}\cdot \dot{\mathbf{y}})}{\partial y} \cdot \overline{\overline{\boldsymbol{M}}} \cdot \overline{\boldsymbol{J}} \cdot \dot{\mathbf{y}}. \tag{4.64}$$

If we now insert (4.64) and (4.63) into (4.61) in compliance with (4.50), the third term on the right side of (4.64) is eliminated. This means that the direct evaluation of Lagrange equations of the second kind in their original form (4.61) involves an unnecessary level of calculational complexity. If we set up the equations of motion according to the d'Alembert principle on the other hand, we reach our goal immediately. For this reason, we will only make use of the d'Alembert and Jourdain principles. These two principles are ultimately based on the partitioning of the space spanned by the coordinates of an exposed mechanical system into two orthogonal subspaces for the free and restricted directions of motion. These orthogonal subspaces, assuming ideal forces (such as appear, for example, in ordinary multibody systems), are mutually independent, which results in uncoupled equations of motion and equations of reaction.

Chapter 5
Multibody Systems

A multibody system consists of rigid bodies between which act internal forces and torques that originate from massless constraint and coupling elements. In addition, other arbitrary external forces and torques can also influence the system. A mass point system is a special case of a multibody system. For example, we can represent a multibody system as a mass point system if all rotational velocities and all internal and external torques vanish with respect to the body center of mass. In comparison to a free multibody system, a free mass point system has only half the number of degrees of freedom because of the omitted rotations. In the case of a planar multibody system, one displacement coordinate and two angular coordinates are dropped as well as one force coordinate and two torque coordinates. Moreover, all bodies must move in parallel principal planes of inertia. In comparison to a free, three-dimensional multibody system, the number of degrees of freedom in a free planar multibody system is reduced by half. Similar simplifications result in gyroscopic systems or planar point systems. In order to limit the diversity of variants, we will only discuss the three-dimensional multibody system. The simplifications in the aforementioned special cases, which lead to a complete cancelation of vanishing equations, will be left to the reader, though they will to some extent be encountered in the examples.

On the basis of the local equations of motion of a free rigid body, the global Newton-Euler equations will be formulated. From these, we can obtain the equations of motion for ideal systems without friction or contact forces, which have a varying form for ordinary and general multibody systems. Also from the Newton-Euler equations follow the equations of reaction, which can be solved for ideal systems independently of the equations of motion. In the context of reaction forces, we will also discuss questions of strength calculation and mass balancing. In non-ideal systems with friction, the equations of motion and equations of reaction are coupled, what demands a common solution of both equation systems. The chapter closes with a few remarks regarding the formalisms available today for setting up equations of motion.

W. Schiehlen and P. Eberhard, *Applied Dynamics*, DOI 10.1007/978-3-319-07335-4_5,

5.1 Local Equations of Motion

The local equations of motion of a multibody system are valid for free bodies. One body K can thus be singled out, see Fig. 5.1, without limiting generality. With the 6×1 position vector $\boldsymbol{x}(t)$ in compliance with (2.81), we can then merge Newton and Euler equations, (3.22) and (3.30), with (2.120) and (2.121),

$$\overline{\overline{\boldsymbol{M}}}(\boldsymbol{x}) \cdot \overline{\boldsymbol{H}}(\boldsymbol{x}) \cdot \ddot{\boldsymbol{x}}(t) + \overline{\boldsymbol{q}}^c(\boldsymbol{x},\dot{\boldsymbol{x}}) = \overline{\boldsymbol{q}}^e(t). \tag{5.1}$$

Here, $\overline{\overline{\boldsymbol{M}}} = \boldsymbol{diag}\{m\boldsymbol{E}\,\boldsymbol{I}\}$ is a symmetric 6×6 block diagonal matrix, $\overline{\boldsymbol{H}} = [\overline{\boldsymbol{H}}_T^T\, \overline{\boldsymbol{H}}_R^T]^T$ a 6×6 functional matrix, and $\overline{\boldsymbol{q}}^c$ a 6×1 vector of the Coriolis and centrifugal forces or the gyroscopic torque. The 6×1 vector $\overline{\boldsymbol{q}}^e = [\boldsymbol{f} \quad \boldsymbol{l}]$, which is also called a wrench, also contains all external or applied forces $\boldsymbol{f}$ and torques $\boldsymbol{l}$ acting on the free body K, while reaction forces and torques do not appear according to definition. In (5.1), the inertia matrix $(\overline{\overline{\boldsymbol{M}}} \cdot \overline{\boldsymbol{H}})$ has become unsymmetrical due to the introduction of the generalized coordinates $\boldsymbol{x}(t)$. However, it can be re-symmetrized by left-multiplication with $\overline{\boldsymbol{H}}^T$. We then have the local equations of motion

$$\boldsymbol{M}(\boldsymbol{x}) \cdot \ddot{\boldsymbol{x}}(t) + \boldsymbol{k}(\boldsymbol{x},\dot{\boldsymbol{x}}) = \boldsymbol{q}(t) \tag{5.2}$$

with the symmetric 6×6 inertia matrix $\boldsymbol{M}$ and the 6×1 vectors $\boldsymbol{k}$ and $\boldsymbol{q}$ of the generalized gyroscopic forces and the generalized applied forces. The local equations of motion of a rigid body are of little practical use, since they cannot be solved themselves because the generalized applied forces generally depend on the position and velocity of the other bodies of the system. Yet the local equations of motion make it easier to understand the global equations of motion of the complete system.

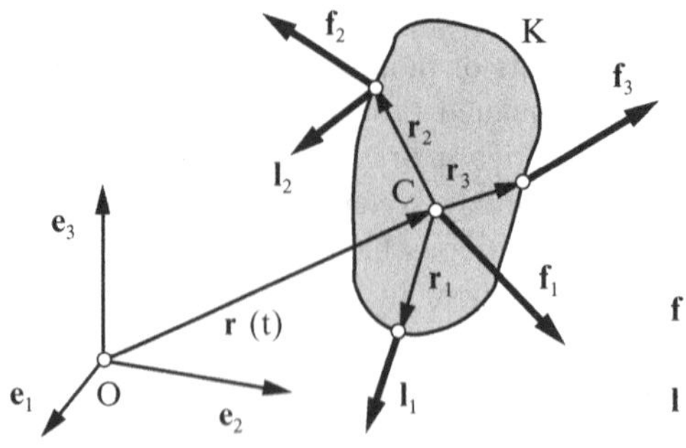

$$\mathbf{f} = \sum_{j=1}^{3} \mathbf{f}_j$$

$$\mathbf{l} = \sum_{j=1}^{3} (\tilde{\mathbf{r}}_j \cdot \mathbf{f}_j + \mathbf{l}_j)$$

Fig. 5.1 Forces and torques acting on a free body K

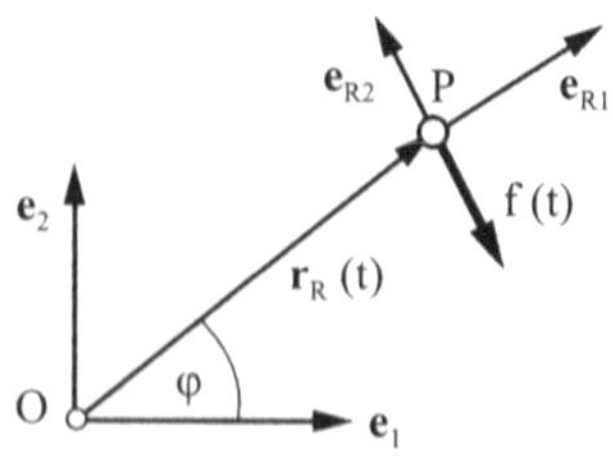

Fig. 5.2 Planar point motion with a reference frame

The Newton and Euler equations in (5.1) refer to the inertial frame. They can, however, also be applied to the moving, but not necessarily body-fixed reference frame R. From (3.58) and (3.59), we then obtain

$$ {}_R\overline{\overline{\boldsymbol{M}}}(\boldsymbol{x}) \cdot {}_R\overline{\boldsymbol{H}}(\boldsymbol{x}_1) \cdot \ddot{\boldsymbol{x}}(t) + {}_R\overline{\boldsymbol{q}}^c(\boldsymbol{x},\dot{\boldsymbol{x}}) = {}_R\overline{\boldsymbol{q}}^e(t) \tag{5.3} $$

with the transformation relations

$$ \overline{\overline{\boldsymbol{M}}} = \overline{\boldsymbol{S}}_R \cdot {}_R\overline{\overline{\boldsymbol{M}}} \cdot \overline{\boldsymbol{S}}_R^T, \tag{5.4} $$

and

$$ \overline{\boldsymbol{H}} = \overline{\boldsymbol{S}}_R \cdot {}_R\overline{\boldsymbol{H}}, \qquad \overline{\boldsymbol{q}}^c = \overline{\boldsymbol{S}}_R \cdot {}_R\overline{\boldsymbol{q}}^c, \qquad \overline{\boldsymbol{q}}^e = \overline{\boldsymbol{S}}_R \cdot {}_R\overline{\boldsymbol{q}}^e. \tag{5.5} $$

These transformations correspond formally to (2.248), but in the present case the transformation matrix $\overline{\boldsymbol{S}}_R$ represents a 6×6 block diagonal matrix,

$$ \overline{\boldsymbol{S}}_R = \boldsymbol{diag}\{\boldsymbol{S}_R \quad \boldsymbol{S}_R\}. \tag{5.6} $$

This means that the Newton equation and the Euler equation depend on the selection of the reference frame. In contrast, the equations of motion (5.2) are invariant to transformations of the frame. Proof of the invariance of the equations of motion can be found by multiplying (5.3) from the left with ${}_R\overline{\boldsymbol{H}}^T$ and then inserting the inverse transformations (5.4) and (5.5). Despite the invariance of the result, is can be advantageous to use a moving reference frame, since the calculation steps used to set up equations of motion in moving reference frames are often simpler.

Example 5.1 (Planar Point Motion). A free mass point P should be impelled by an applied force $f(t)$ which acts perpendicularly to the distance vector between the frame origin O and the point P, see Fig. 5.2. The equations of motion are set up in the coordinates of the inertial frame I and the moving reference frame R. The polar coordinates serve as generalized coordinates

$$ \boldsymbol{x}(t) = \left[\, r \quad \varphi \,\right]. \tag{5.7} $$

The Newton equations (5.2) are written in the inertial frame

$$\begin{bmatrix} m & 0 \\ 0 & m \end{bmatrix} \cdot \begin{bmatrix} \cos\varphi & -r\sin\varphi \\ \sin\varphi & r\cos\varphi \end{bmatrix} \cdot \begin{bmatrix} \ddot{r} \\ \ddot{\varphi} \end{bmatrix} + m\begin{bmatrix} -2\dot{r}\dot{\varphi}\sin\varphi - r\dot{\varphi}^2\cos\varphi \\ +2\dot{r}\dot{\varphi}\cos\varphi - r\dot{\varphi}^2\sin\varphi \end{bmatrix} = \begin{bmatrix} -\sin\varphi \\ \cos\varphi \end{bmatrix} f(t). \tag{5.8}$$

By left-multiplication of (5.8) with $\overline{\boldsymbol{H}}_T^T$, we then find the equations of motion

$$\begin{bmatrix} m & 0 \\ 0 & mr^2 \end{bmatrix} \cdot \begin{bmatrix} \ddot{r} \\ \ddot{\varphi} \end{bmatrix} + \begin{bmatrix} -mr\dot{\varphi}^2 \\ 2mr\dot{r}\dot{\varphi} \end{bmatrix} = \begin{bmatrix} 0 \\ rf(t) \end{bmatrix}. \tag{5.9}$$

The reference frame is given by

$$_R\boldsymbol{r}_R(\boldsymbol{x}) = \begin{bmatrix} r \\ 0 \end{bmatrix}, \qquad \boldsymbol{S}_R(\boldsymbol{x}) = \begin{bmatrix} \cos\varphi & -\sin\varphi \\ \sin\varphi & \cos\varphi \end{bmatrix} \tag{5.10}$$

and for the relative position vector

$$\boldsymbol{r}_R = \boldsymbol{0} \tag{5.11}$$

applies. We obtain the Newton equations in the reference frame in compliance with (2.257)

$$\begin{bmatrix} m & 0 \\ 0 & m \end{bmatrix} \cdot \begin{bmatrix} 1 & 0 \\ 0 & r \end{bmatrix} \cdot \begin{bmatrix} \ddot{r} \\ \ddot{\varphi} \end{bmatrix} + \begin{bmatrix} -mr\dot{\varphi}^2 \\ 2m\dot{r}\dot{\varphi} \end{bmatrix} = \begin{bmatrix} 0 \\ f(t) \end{bmatrix}. \tag{5.12}$$

We can see the much simpler structure of the Newton equation (5.12) in comparison to (5.8). By left-multiplication with $_R\boldsymbol{H}_T^T$, we also obtain from (5.12) the equations of motion (5.9) in unaltered form.

This example confirms the insight that moving reference frames simplify the calculation steps used to set up equations of motion without affecting the outcome themselves.

End of Example 5.1.

The inertia matrix $\boldsymbol{M}$ of the local equations of motion (5.2) can assume the form of a block diagonal matrix given the appropriate choice of generalized coordinates. This is particularly the case when the coordinates of the 3×1 location vector $\boldsymbol{r}$ are used for the center of mass C with respect to the inertial frame. With the 6×1 position vector

$$\boldsymbol{x}(t) = \begin{bmatrix} r_1 & r_2 & r_3 & \alpha & \beta & \gamma \end{bmatrix} \tag{5.13}$$

we obtain the 6×6 inertia matrix

$$\boldsymbol{M} = \left[\begin{array}{c|c} m\boldsymbol{E} & \boldsymbol{0} \\ \hline \boldsymbol{0} & \boldsymbol{H}_R^T \cdot \boldsymbol{I} \cdot \boldsymbol{H}_R \end{array}\right] \tag{5.14}$$

where the 3×3 Jacobian matrix of the Cardano angles is given by (2.100). If on the other hand we use the coordinates of the 3×1 location vector $\boldsymbol{r}_K = \boldsymbol{r}_C + \boldsymbol{r}_{CK}$ of a node as generalized coordinates, then the 6×1 position vector is

$$\boldsymbol{x}'(t) = \left[\, r_{1K} \;\; r_{2K} \;\; r_{3K} \;\; \alpha \;\; \beta \;\; \gamma \,\right] \tag{5.15}$$

and the associated 6×6 inertia matrix

$$\boldsymbol{M}' = \left[\begin{array}{c|c} m\boldsymbol{E} & m\tilde{\boldsymbol{r}}_{CK} \cdot \boldsymbol{H}_R \\ \hline \boldsymbol{H}_R^T \cdot \tilde{\boldsymbol{r}}_{CK}^T m & \boldsymbol{H}_R^T \cdot \boldsymbol{I}' \cdot \boldsymbol{H}_R \end{array}\right] \tag{5.16}$$

is fully occupied. When calculating the inertia matrix (5.16), it should be noted that $\boldsymbol{r}_{CK}$ = constant applies in a body-fixed frame. The inertia matrix (5.14) makes the fact clear that the rotation and translation of a free rigid body K, assuming corresponding external forces and torques, are decoupled with respect to the center of mass C. This is not the case for an arbitrary moving reference point (e.g. a node), as (5.16) directly shows.

5.2 Newton-Euler Equations

If we now write the Newton and Euler equations for every body $K_i, i = 1(1)p$ of a multibody system, we then obtain the global system equations, also called Newton-Euler equations. It is now necessary to distinguish the type of constraints in accordance with Chap. 2.

For free systems with the $e \times 1$ position vector $\boldsymbol{x}(t)$, the Newton-Euler equations reads as

$$\overline{\overline{\boldsymbol{M}}}(\boldsymbol{x}) \cdot \overline{\boldsymbol{H}}(\boldsymbol{x}) \cdot \ddot{\boldsymbol{x}}(t) + \overline{\boldsymbol{q}}^c(\boldsymbol{x}, \dot{\boldsymbol{x}}) = \overline{\boldsymbol{q}}^e(t). \tag{5.17}$$

In holonomic systems, there are also reaction forces in play which according to (3.41) can be expressed by the $q \times 1$ vector $\boldsymbol{g}(t)$ of the generalized reaction forces. We then obtain from (3.35) with the $f \times 1$ position vector $\boldsymbol{y}(t)$ the Newton-Euler equations in the form

$$\overline{\overline{\boldsymbol{M}}}(\boldsymbol{y}, t) \cdot \overline{\boldsymbol{J}}(\boldsymbol{y}, t) \cdot \ddot{\boldsymbol{y}}(t) + \overline{\boldsymbol{q}}^c(\boldsymbol{y}, \dot{\boldsymbol{y}}, t) = \overline{\boldsymbol{q}}^e(t) + \overline{\boldsymbol{Q}} \cdot \boldsymbol{g}(t). \tag{5.18}$$

Additional constraints are added in nonholonomic systems, so the $(q+r) \times 1$ vector $\boldsymbol{g}(t)$ of the generalized reaction forces appears. With the $g \times 1$ velocity vector $\boldsymbol{z}(t)$, we can then write the Newton-Euler equations in the form

$$\overline{\overline{\boldsymbol{M}}}(y,t) \cdot \overline{\boldsymbol{L}}(y,z,t) \cdot \dot{z}(t) + \overline{\boldsymbol{q}}^c(y,z,t) = \overline{\boldsymbol{q}}^e(t) + \overline{\boldsymbol{Q}} \cdot \boldsymbol{g}(t). \tag{5.19}$$

Here, the $e \times e$ block diagonal matrix reads in all three cases

$$\overline{\overline{\boldsymbol{M}}} = \boldsymbol{diag}\{m_1\boldsymbol{E} \quad m_2\boldsymbol{E} \ \dots \ m_p\boldsymbol{E} \ \boldsymbol{I}_1 \ \dots \ \boldsymbol{I}_p\}, \tag{5.20}$$

containing the mass and inertia tensors. The global functional or Jacobian matrices $\overline{\boldsymbol{H}}, \overline{\boldsymbol{J}}, \overline{\boldsymbol{L}}$ are written as

$$\overline{\boldsymbol{H}} = \begin{bmatrix} \boldsymbol{H}_{T1}^T & \boldsymbol{H}_{T2}^T & \dots & \boldsymbol{H}_{Tp}^T & \boldsymbol{H}_{R1}^T & \dots & \boldsymbol{H}_{Rp}^T \end{bmatrix}^T, \tag{5.21}$$

the global distribution matrix of the reaction forces is

$$\overline{\boldsymbol{Q}} = \begin{bmatrix} \boldsymbol{F}_1^T & \boldsymbol{F}_2^T & \dots & \boldsymbol{F}_p^T & \boldsymbol{L}_1^T & \dots & \boldsymbol{L}_p^T \end{bmatrix}^T, \tag{5.22}$$

and for the global $e \times 1$ force vectors $\overline{\boldsymbol{q}}^c$ and $\overline{\boldsymbol{q}}^e$ there applies, respectively,

$$\overline{\boldsymbol{q}} = \begin{bmatrix} \boldsymbol{f}_1 & \boldsymbol{f}_2 & \dots & \boldsymbol{f}_p & \boldsymbol{l}_1 & \dots & \boldsymbol{l}_p \end{bmatrix}. \tag{5.23}$$

The Newton-Euler equations (5.17)–(5.19) each represent $6p$ scalar equations. With them, $6p$ unknowns can be determined at a time. Motions and/or forces appear as unknowns. In a free system, the applied forces are given, so all motions can be determined, which is known as a direct problem. In the case of an indirect or inverse problem on the other hand, all motions are defined by rheonomic constraints, and the reaction forces are sought. Yet a combination of these cases, which is called a mixed problem, is also possible. Holonomic and nonholonomic systems belong to the category of mixed problems since some of the motions are given by constraints. For example, in a holonomic system f motions can be calculated, and there are an additional q reaction forces. In total, there are thus $f+q$ unknowns. An indirect problem is defined as a system with $f = 0$ degrees of freedom and q reaction forces, which represents a statically determinate multibody system in the case of scleronomic constraints. On the other hand, if $q > 6p$ reaction forces are unknown, then the system is statically indeterminate. The reaction forces cannot be uniquely calculated in statically indeterminate systems. They are open to calculation only if the reaction forces are first substituted with applied forces to a sufficient extent, e.g. by replacing constraints by elastic elements.

The solution of the Newton-Euler equations (5.18) and (5.19) of constrained systems is not trivial. As a result of the mixed problem, (5.18) and (5.19) are no longer pure differential equations, but rather coupled differential-algebraic equations. These can be solved directly with complex numerical methods, see e.g.

Eich-Soellner and Führer [18] or Simeon [59]. On the other hand, the principles of mechanics allow an extensive if not complete decoupling of the mixed problem, leaving us with separately solvable equations of motion and reaction.

In order to solve directly, we waive the generalized coordinates completely, so the inertia matrix retains the block diagonal form

$$\overline{\overline{\boldsymbol{M}}} \cdot \ddot{\boldsymbol{x}}(t) + \overline{\boldsymbol{q}}^c(\boldsymbol{x},\dot{\boldsymbol{x}}) = \overline{\boldsymbol{q}}^e(t) + \overline{\boldsymbol{Q}} \cdot \boldsymbol{g}(t). \tag{5.24}$$

If we also note that the distribution matrix $\overline{\boldsymbol{Q}}$ can be determined by partial differentiation from the implicit constraints (2.175)

$$\overline{\boldsymbol{Q}}^T = \frac{\partial \boldsymbol{\phi}}{\partial \boldsymbol{x}} = \boldsymbol{\phi}_x, \tag{5.25}$$

as was shown in Sect. 4.1, and since the generalized reaction forces can be interpreted as Lagrange multipliers, we then obtain after a twofold total differentiation of (2.175) in compliance with (4.55) with (5.24),

$$\begin{bmatrix} \overline{\overline{\boldsymbol{M}}} & -\boldsymbol{\phi}_x^T \\ -\boldsymbol{\phi}_x & \boldsymbol{0} \end{bmatrix} \cdot \begin{bmatrix} \ddot{\boldsymbol{x}} \\ \boldsymbol{g} \end{bmatrix} = \begin{bmatrix} \overline{\boldsymbol{q}}^e - \boldsymbol{q}^c \\ \dot{\boldsymbol{\phi}}_t + \dot{\boldsymbol{\phi}}_x \cdot \dot{\boldsymbol{x}} \end{bmatrix}. \tag{5.26}$$

As a result of the twofold differentiation of the constraints (2.175), the differential-algebraic system of Eq. (5.26) has a double zero eigenvalue and is thus numerically unstable. The integration method for differential-algebraic equations ensures automatic stabilization of the system (5.26). For the generalized reaction forces, we then find the relation

$$\boldsymbol{g} = (\boldsymbol{\phi}_x \cdot \overline{\overline{\boldsymbol{M}}}^{-1} \cdot \boldsymbol{\phi}_x^T)^{-1} \cdot \left[\boldsymbol{\phi}_x \cdot \overline{\overline{\boldsymbol{M}}}^{-1} \cdot (\overline{\boldsymbol{q}}^c - \overline{\boldsymbol{q}}^e) - \dot{\boldsymbol{\phi}}_t - \dot{\boldsymbol{\phi}}_x \cdot \dot{\boldsymbol{x}}\right]. \tag{5.27}$$

With a third total differentiation of (5.27), the differential-algebraic system of equations can be converted into a pure differential system of equations. The total number of total differentiations is also called the index. Multibody systems represent differential-algebraic systems of index 3. The initial conditions of integration $\boldsymbol{x}_0$, $\dot{\boldsymbol{x}}_0$, and $\boldsymbol{g}_0$, must be satisfied by Eqs. (5.26) and (5.27), i.e. the constraints and their derivatives, which is a further difficulty in the numerical solution.

5.3 Equations of Motion of Ideal Systems

An ideal system is characterized by the fact that the applied forces do not depend on the reaction forces. This is the case, for example, when all contact and friction forces disappear. If all the applied forces exhibit proportional-differential behavior and only holonomic constraints appear, it is considered as an ordinary multibody system.

Ordinary multibody systems, from a mathematical point of view, are characterized by the fact that the equations of motion can be transformed into a pure vector differential equation of the second order. All multibody systems, also those that are not ordinary, are called general multibody systems. These include in particular the nonholonomic systems and systems with proportional-integral applied forces.

5.3.1 Ordinary Multibody Systems

The equations of motion of a free multibody system with only proportional-differential forces are obtained from (5.17) with (3.11) by left-multiplication with $\overline{\boldsymbol{H}}^T$ in the form

$$\boldsymbol{M}(\boldsymbol{x}) \cdot \ddot{\boldsymbol{x}}(t) + \boldsymbol{k}(\boldsymbol{x},\dot{\boldsymbol{x}}) = \boldsymbol{q}(\boldsymbol{x},\dot{\boldsymbol{x}},t), \tag{5.28}$$

where

$$\boldsymbol{M}(\boldsymbol{x}) = \overline{\boldsymbol{H}}^T \cdot \overline{\overline{\boldsymbol{M}}} \cdot \overline{\boldsymbol{H}} \tag{5.29}$$

is the symmetric $e \times e$ inertia matrix, $\boldsymbol{k}(\boldsymbol{x},\dot{\boldsymbol{x}})$ the $e \times 1$ vector of the generalized gyroscopic forces, and $\boldsymbol{q}(\boldsymbol{x},\dot{\boldsymbol{x}},t)$ the $e \times 1$ vector of the generalized applied forces. The generalized gyroscopic forces are thus based on the Coriolis and centrifugal forces as well as the gyroscopic torques in the Newton-Euler equations.

However, the global equations of motion (5.28) of a free system can also be found by merging the local equations of motion (5.2). For example, if we use the global $e \times 1$ position vector

$$\boldsymbol{x}(t) = \begin{bmatrix} \boldsymbol{x}_1 & \boldsymbol{x}_2 & \dots & \boldsymbol{x}_p \end{bmatrix}, \tag{5.30}$$

which is comprised of the local 6×1 position vectors $\boldsymbol{x}_i(t), i = 1(1)p$, we then obtain for the inertia matrix and force vectors of the global equations of motion

$$\boldsymbol{M}(\boldsymbol{x}) = \boldsymbol{diag}\{\boldsymbol{M}_1\, \boldsymbol{M}_2\, \dots\, \boldsymbol{M}_p\}, \tag{5.31}$$

$$\boldsymbol{k}(\boldsymbol{x},\dot{\boldsymbol{x}}) = \begin{bmatrix} \boldsymbol{k}_1 & \boldsymbol{k}_2 & \dots & \boldsymbol{k}_p \end{bmatrix}, \tag{5.32}$$

$$\boldsymbol{q}(\boldsymbol{x},\dot{\boldsymbol{x}},t) = \begin{bmatrix} \boldsymbol{q}_1 & \boldsymbol{q}_2 & \dots & \boldsymbol{q}_p \end{bmatrix}. \tag{5.33}$$

The proof is easily demonstrated. With (5.21), the global $e \times e$ functional matrix takes on the special form

$$\overline{\boldsymbol{H}} = \boldsymbol{diag}\{\overline{\boldsymbol{H}}_1 \quad \overline{\boldsymbol{H}}_2 \ \dots \ \overline{\boldsymbol{H}}_p\} \tag{5.34}$$

where the local 6×6 functional matrices $\overline{\boldsymbol{H}}_i$ appear, see (5.1).

Fig. 5.3 Applied forces of an one-dimensional slipping coupling element

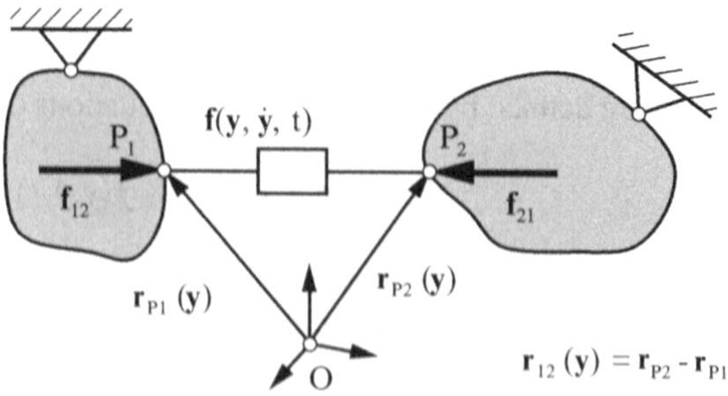

As opposed to the local equations of motion (5.2), the global equations of motion (5.28) represent a complete system of differential equations, since now the position and velocity of all bodies are available. The global equations of motion can be solved for given initial conditions $\boldsymbol{x}(t_0) = \boldsymbol{x}_0$, $\dot{\boldsymbol{x}}(t_0) = \dot{\boldsymbol{x}}_0$ by integration.

The equations of motion of a holonomic multibody system with proportional-differential forces are found from (5.18) with (3.11) using the d'Alembert principle (4.29) in the form

$$\boldsymbol{M}(\boldsymbol{y},t)\cdot\ddot{\boldsymbol{y}}(t) + \boldsymbol{k}(\boldsymbol{y},\dot{\boldsymbol{y}},t) = \boldsymbol{q}(\boldsymbol{y},\dot{\boldsymbol{y}},t). \tag{5.35}$$

According to the d'Alembert principle, the reaction forces are dropped in compliance with the orthogonality relation (4.16). Moreover, a symmetrization of the $f \times f$ inertia matrix takes place

$$\boldsymbol{M}(\boldsymbol{y},t) = \overline{\boldsymbol{J}}^T \cdot \overline{\overline{\boldsymbol{M}}} \cdot \overline{\boldsymbol{J}}. \tag{5.36}$$

The mass matrices (5.29) and (5.36) are marked with the same letters although they have different dimensions and of course different elements. This assignment is however perfectly clear in connection with the equations of motion, so for the sake of improved legibility we do not use indices for these matrices. The same is true for the $f \times 1$ vectors $\boldsymbol{k}$ and $\boldsymbol{q}$ of the generalized gyroscopic forces and of the generalized applied forces.

If the applied forces can be traced back to one-dimensional sliding coupling elements, which are applied between two points P_1 and P_2 of different rigid bodies. Then, according to Fig. 5.3 it follows

$$\boldsymbol{f}_{12} = -\boldsymbol{f}_{21} = \frac{\boldsymbol{r}_{12}(\boldsymbol{y})}{\sqrt{\boldsymbol{r}_{12}\cdot\boldsymbol{r}_{12}}} f(\boldsymbol{y},\dot{\boldsymbol{y}},t). \tag{5.37}$$

The function f is the scalar force law typical of the coupling element under investigation, e.g. $f = -c(r_{12} - L)$ with the tension-free spring length L and the current length $r_{12} = \sqrt{\boldsymbol{r}_{12}\cdot\boldsymbol{r}_{12}}$. According to the d'Alembert principle (4.24), however, we can also calculate the generalized applied coupling forces directly from

$$\boldsymbol{q} = \frac{\partial r_{12}}{\partial \boldsymbol{y}} f(\boldsymbol{y},\dot{\boldsymbol{y}},t). \tag{5.38}$$

This relation is useful when the scalar distance $r_{12}(\mathbf{y})$ can be stated in a simple way.

The equations of motion (5.28) and (5.35) describe ordinary multibody systems, which are defined by vector differential equations of the form

$$\boldsymbol{M}(\mathbf{y},t)\cdot\ddot{\mathbf{y}}(t)+\boldsymbol{f}(\mathbf{y},\dot{\mathbf{y}},t)=\mathbf{0} \tag{5.39}$$

with positive definite inertia matrix $\boldsymbol{M}(\mathbf{y},t)$. Ordinary multibody systems are thus characterized by holonomic constraints and ideal, proportional-differential forces.

The equations of motion (5.35) can be obtained not only from the Newton-Euler equations (5.18), but also from the equations of motion (5.28) of the free system. Based on the explict constraint equations from (2.176) and the corresponding Jacobian matrix $\boldsymbol{I}$ as,

$$\boldsymbol{x}=\boldsymbol{x}(\mathbf{y},t), \qquad \dot{\boldsymbol{x}}=\boldsymbol{I}(\mathbf{y},t)\cdot\dot{\mathbf{y}}(t)+\frac{\partial \boldsymbol{x}}{\partial t} \tag{5.40}$$

the following relations apply

$$\boldsymbol{M}_{\text{holonomic}}(\mathbf{y},t)=\boldsymbol{I}^T\cdot\boldsymbol{M}_{\text{free}}(\boldsymbol{x},t)\cdot\boldsymbol{I}, \tag{5.41}$$

$$\boldsymbol{k}_{\text{holonomic}}(\mathbf{y},\dot{\mathbf{y}},t)=\boldsymbol{I}^T\cdot\left[\boldsymbol{M}_{\text{free}}(\boldsymbol{x},t)\cdot\dot{\boldsymbol{I}}\cdot\dot{\mathbf{y}}+\boldsymbol{M}_{\text{free}}(\boldsymbol{x},t)\cdot\frac{\partial^2\boldsymbol{x}}{\partial t^2}+\boldsymbol{k}_{\text{free}}(\boldsymbol{x},\dot{\boldsymbol{x}})\right], \tag{5.42}$$

$$\boldsymbol{q}_{\text{holonomic}}(\mathbf{y},\dot{\mathbf{y}},t)=\boldsymbol{I}^T\cdot\boldsymbol{q}_{\text{free}}(\boldsymbol{x},\dot{\boldsymbol{x}},t), \tag{5.43}$$

where the arguments on the right-hand side should be substituted with (5.40). The reaction forces and torques which additionally appear due to the constraints are dropped again in the relations (5.41)–(5.43) in accordance with (4.17). Thus, if only the equations of motion are being sought, we can fully dispense with adding the reaction forces, as is the case with the Lagrange equations of the second kind. Corresponding relations also apply if additional holonomic constraints are imposed on the holonomic system.

Example 5.2 (Physical Pendulum). The equations of motion of a physical pendulum will now be obtained from the equations of motion of the associated double pendulum, see Fig. 5.4.

Taking into account the position vector (2.180) and the vector (3.14) of the generalized reaction forces, the Newton-Euler equations (5.18) of the double pendulum are

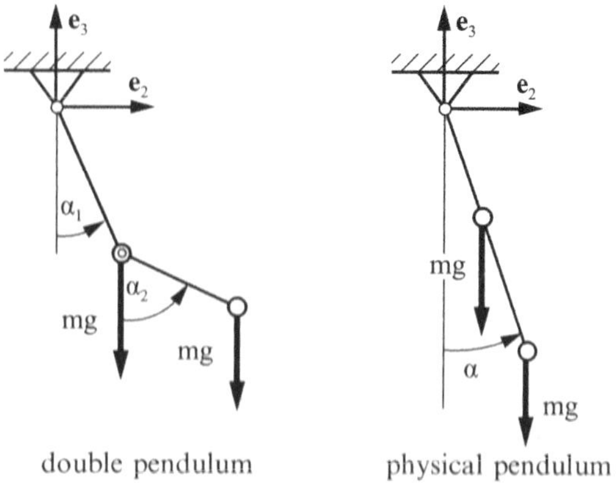

Fig. 5.4 Transition from a double pendulum to a physical pendulum

$$\underbrace{mL\begin{bmatrix} 0 & 0 \\ \cos\alpha_1 & 0 \\ \sin\alpha_1 & 0 \\ 0 & 0 \\ \cos\alpha_1 & \cos\alpha_2 \\ \sin\alpha_1 & \sin\alpha_2 \end{bmatrix}}_{\overline{\overline{\boldsymbol{M}}}\cdot\boldsymbol{J}} \cdot \underbrace{\begin{bmatrix} \ddot{\alpha}_1 \\ \ddot{\alpha}_2 \end{bmatrix}}_{\ddot{\boldsymbol{y}}(t)} + \underbrace{mL\begin{bmatrix} 0 \\ -\dot{\alpha}_1^2\sin\alpha_1 \\ -\dot{\alpha}_1^2\cos\alpha_1 \\ 0 \\ -\dot{\alpha}_1^2\sin\alpha_1 - \dot{\alpha}_2^2\sin\alpha_2 \\ \dot{\alpha}_1^2\cos\alpha_1 + \dot{\alpha}_2^2\cos\alpha_2 \end{bmatrix}}_{\overline{\boldsymbol{q}}^c(\boldsymbol{y},\dot{\boldsymbol{y}})}$$

$$= \underbrace{\begin{bmatrix} 0 \\ 0 \\ -mg \\ 0 \\ 0 \\ -mg \end{bmatrix}}_{\overline{\boldsymbol{q}}^e(\boldsymbol{y})} + \underbrace{\begin{bmatrix} 1 & 0 & 0 & 0 \\ 0 & -\sin\alpha_1 & 0 & \sin\alpha_2 \\ 0 & 0 & 0 & -\cos\alpha_2 \\ 0 & 0 & 1 & 0 \\ 0 & 0 & 0 & -\sin\alpha_2 \\ 0 & 0 & 0 & \cos\alpha_2 \end{bmatrix}}_{\overline{\boldsymbol{Q}}(\boldsymbol{y})} \cdot \underbrace{\begin{bmatrix} g_1 \\ g_2 \\ g_3 \\ g_4 \end{bmatrix}}_{\boldsymbol{g}(t)}. \tag{5.44}$$

We then find for the equations of motion of the double pendulum in accordance with the d'Alembert principle

$$\underbrace{mL^2\begin{bmatrix} 2 & \cos(\alpha_1-\alpha_2) \\ \cos(\alpha_1-\alpha_2) & 1 \end{bmatrix}}_{\boldsymbol{M}(\boldsymbol{y})} \cdot \underbrace{\begin{bmatrix} \ddot{\alpha}_1 \\ \ddot{\alpha}_2 \end{bmatrix}}_{\ddot{\boldsymbol{y}}(t)}$$

$$+\underbrace{mL^2\begin{bmatrix} \dot{\alpha}_2^2\sin(\alpha_1-\alpha_2) \\ -\dot{\alpha}_1^2\sin(\alpha_1-\alpha_2) \end{bmatrix}}_{\boldsymbol{k}(\boldsymbol{y},\dot{\boldsymbol{y}})} = \underbrace{-mgL\begin{bmatrix} 2\sin\alpha_1 \\ \sin\alpha_2 \end{bmatrix}}_{\boldsymbol{q}(\boldsymbol{y})}. \tag{5.45}$$

Now the additional constraint $\alpha_1 - \alpha_2 = 0$ is introduced or

$$\boldsymbol{y} = \begin{bmatrix} 1 \\ 1 \end{bmatrix} \alpha, \tag{5.46}$$

giving rise to an additional reaction torque, which is not written however. Equation (5.46) yields the 2×1 Jacobian matrix

$$\boldsymbol{I} = \begin{bmatrix} 1 & 1 \end{bmatrix} \tag{5.47}$$

and with the relations (5.41) we obtain the scalar equations of motion of the physical pendulum

$$5mL^2 \ddot{\alpha}(t) = -3mgL \sin \alpha(t). \tag{5.48}$$

The corresponding nonlinear vibration equation

$$\ddot{\alpha}(t) + \nu^2 \sin \alpha(t) = 0, \qquad \nu^2 = \frac{3}{5} \frac{g}{L} \tag{5.49}$$

can be solved in an analytical closed form, see e.g. Magnus and Müller-Slany [36].

End of Example 5.2.

In practice, not only the general nonlinear equations of motion (5.35) but also the linearized equations of motion play an important role. First, we will linearize the equation of motion with reference to a target motion. Then some information will be provided regarding the computation of linearized equations.

The target motion of a mechanical system can either be grounded in the system itself or given by the engineering task. Characteristic target motions of a system are its particular solutions

$$\boldsymbol{y}_S(t) = \boldsymbol{y}_P(t) \tag{5.50}$$

with

$$\boldsymbol{M}(\boldsymbol{y}_P, t) \cdot \ddot{\boldsymbol{y}}_p(t) + \boldsymbol{k}(\boldsymbol{y}_P, \dot{\boldsymbol{y}}_p, t) = \boldsymbol{q}(\boldsymbol{y}_P, \dot{\boldsymbol{y}}_p, t), \tag{5.51}$$

to which also belong the zero position or other positions of equilibrium, and given motions $\boldsymbol{y}_S(t)$ provided by the engineering problem.

In the vicinity of the target motions, the system should carry out a small neighboring motion. Then, as shown in (2.278), it yields

$$\boldsymbol{y}(t) = \boldsymbol{y}_S(t) + \boldsymbol{\eta}(t), \qquad ||\boldsymbol{\eta}(t)|| \ll a, \tag{5.52}$$

where $\boldsymbol{y}(t)$ represents the $f \times 1$ position vector of large motions, $\boldsymbol{y}_S(t)$ describes the $f \times 1$ vector of the target motion, and $\boldsymbol{\eta}(t)$ is now introduced as the $f \times 1$ position vector of small motions. Here, a is a problem-specific reference value. Analogously, the velocities and accelerations can also be linearized, i.e.,

$$\dot{\boldsymbol{y}}(t) = \dot{\boldsymbol{y}}_S(t) + \dot{\boldsymbol{\eta}}(t), \qquad ||\dot{\boldsymbol{\eta}}(t)|| \ll b, \tag{5.53}$$

$$\ddot{\boldsymbol{y}}(t) = \ddot{\boldsymbol{y}}_S(t) + \ddot{\boldsymbol{\eta}}(t), \qquad ||\ddot{\boldsymbol{\eta}}(t)|| \ll c. \tag{5.54}$$

If we now insert (5.52)–(5.54) into (5.35), a Taylor series expansion, in each case up to the first term, yields

$$\boldsymbol{M}(\boldsymbol{y},t) = \boldsymbol{M}_0(t) + \frac{\partial \boldsymbol{M}}{\partial \boldsymbol{y}} \cdot \boldsymbol{\eta}(t), \tag{5.55}$$

$$\boldsymbol{k}(\boldsymbol{y},\dot{\boldsymbol{y}},t) = \boldsymbol{k}_0(t) + \frac{\partial \boldsymbol{k}}{\partial \boldsymbol{y}} \cdot \boldsymbol{\eta}(t) + \frac{\partial \boldsymbol{k}}{\partial \dot{\boldsymbol{y}}} \cdot \dot{\boldsymbol{\eta}}(t), \tag{5.56}$$

$$\boldsymbol{q}(\boldsymbol{y},\dot{\boldsymbol{y}},t) = \boldsymbol{q}_0(t) + \frac{\partial \boldsymbol{q}}{\partial \boldsymbol{y}} \cdot \boldsymbol{\eta}(t) + \frac{\partial \boldsymbol{q}}{\partial \dot{\boldsymbol{y}}} \cdot \dot{\boldsymbol{\eta}}(t). \tag{5.57}$$

From (5.35), neglecting all second-order terms in $\boldsymbol{\eta}$, $\dot{\boldsymbol{\eta}}$, and $\ddot{\boldsymbol{\eta}}$, follow the linearized equations of motion

$$\boldsymbol{M}(t) \cdot \ddot{\boldsymbol{\eta}}(t) + \boldsymbol{P}(t) \cdot \dot{\boldsymbol{\eta}}(t) + \boldsymbol{Q}(t) \cdot \boldsymbol{\eta}(t) = \boldsymbol{h}(t). \tag{5.58}$$

The following abbreviations apply,

$$\begin{aligned}
\boldsymbol{M}(t) &= \boldsymbol{M}_0(t), \\
\boldsymbol{P}(t) &= \frac{\partial \boldsymbol{k}}{\partial \dot{\boldsymbol{y}}} - \frac{\partial \boldsymbol{q}}{\partial \dot{\boldsymbol{y}}}, \\
\boldsymbol{Q}(t) &= \frac{\partial \boldsymbol{M}}{\partial \boldsymbol{y}} \cdot \ddot{\boldsymbol{y}}_S(t) + \frac{\partial \boldsymbol{k}}{\partial \boldsymbol{y}} - \frac{\partial \boldsymbol{q}}{\partial \boldsymbol{y}}, \\
\boldsymbol{h}(t) &= \boldsymbol{q}_0(t) - \boldsymbol{k}_0(t) - \boldsymbol{M}_0(t) \cdot \ddot{\boldsymbol{y}}_S(t).
\end{aligned} \tag{5.59}$$

The term $\boldsymbol{\eta} \cdot \ddot{\boldsymbol{\eta}}$ is also neglected as a quadratically small quantity. In (5.58) we find, in addition to the $f \times f$ inertia matrix $\boldsymbol{M}(t)$, the $f \times f$ matrix $\boldsymbol{P}(t)$ of the velocity-dependent forces and the $f \times f$ matrix $\boldsymbol{Q}(t)$ of the position-dependent forces as well as the $f \times 1$ vector $\boldsymbol{h}(t)$ of the excitation function. If the system is linearized with respect to a particular solution (5.50), then the excitation function vanishes, $\boldsymbol{h}(t) \equiv \boldsymbol{0}$, i.e. it is now a homogeneous time-variant system. On the other hand, we can also have a time-invariant system with constant matrices $\boldsymbol{M}$, $\boldsymbol{P}$, $\boldsymbol{Q}$. Then it is possible to subdivide the matrices into symmetric and skew-symmetric components

$$\boldsymbol{M} \cdot \ddot{\boldsymbol{\eta}}(t) + (\boldsymbol{D} + \boldsymbol{G}) \cdot \dot{\boldsymbol{\eta}}(t) + (\boldsymbol{K} + \boldsymbol{N}) \cdot \boldsymbol{\eta}(t) = \boldsymbol{h}(t) \tag{5.60}$$

with $\boldsymbol{M} = \boldsymbol{M}^T$, $\boldsymbol{D} = \boldsymbol{D}^T$, $\boldsymbol{K} = \boldsymbol{K}^T$ and $\boldsymbol{G} = -\boldsymbol{G}^T$, $\boldsymbol{N} = -\boldsymbol{N}^T$. By multiplying with $\dot{\boldsymbol{\eta}}(t)$ we obtain from (5.60) the time derivative of an energy expression

$$\underbrace{\dot{\boldsymbol{\eta}} \cdot \boldsymbol{M} \cdot \ddot{\boldsymbol{\eta}}}_{\frac{d}{dt}T} + \underbrace{\dot{\boldsymbol{\eta}} \cdot \boldsymbol{D} \cdot \dot{\boldsymbol{\eta}}}_{2R} + \underbrace{\dot{\boldsymbol{\eta}} \cdot \boldsymbol{G} \cdot \dot{\boldsymbol{\eta}}}_{0} + \underbrace{\dot{\boldsymbol{\eta}} \cdot \boldsymbol{K} \cdot \boldsymbol{\eta}}_{\frac{d}{dt}U} + \dot{\boldsymbol{\eta}} \cdot \boldsymbol{N} \cdot \boldsymbol{\eta} = \dot{\boldsymbol{\eta}} \cdot \boldsymbol{h}, \tag{5.61}$$

which allows a physical explanation of the individual terms. The inertia matrix $\boldsymbol{M}$ determines the kinetic energy T and thus the inertia forces, the damping matrix $\boldsymbol{D}$ denotes, via the Rayleigh dissipation function $R > 0$, the damping forces, and the gyroscopic matrix $\boldsymbol{G}$ describes the gyroscopic forces, which do not bring in any change to the energy balance. The stiffness matrix $\boldsymbol{K}$ determines the potential energy U and thus the conservative position forces, while the matrix $\boldsymbol{N}$ summarizes the circulatory forces, also called non-conservative position forces. Also, the energy of the system is affected by the excitation forces $\boldsymbol{h}(t)$. For $\boldsymbol{D} = \boldsymbol{0}$, $\boldsymbol{N} = \boldsymbol{0}$, and $\boldsymbol{h} = \boldsymbol{0}$ the multibody system is conservative, i.e. its total energy is constant,

$$\frac{d}{dt}(T+U) = 0 \qquad \rightarrow \qquad T+U = \text{constant}. \tag{5.62}$$

The nonlinear equations of motion (5.35) are only linearized in practice in the case of simple systems or manual calculations. For implementation in computer programs on the other hand, it is more advantageous to carry out the linearization already at the kinematics step as in Sect. 2.5 and then to set up equations of motion with the linearized kinematic quantities. This simplifies the computation work considerably, and we also obtain the equations of motion in the form (5.58) or (5.60). But it must be noted that the Taylor series expansion of the location vectors $\boldsymbol{r}_i(\boldsymbol{y},t)$ and the rotation tensors $\boldsymbol{S}_i(\boldsymbol{y},t)$ has to be carried out *to the second term*. To build the Jacobian matrices $\boldsymbol{J}_{Ti}(\boldsymbol{y},t)$ and $\boldsymbol{J}_{Ri}(\boldsymbol{y},t)$, a differentiation with respect to the position vector $\boldsymbol{y}(t)$ is required in compliance with (2.284), by which the occurring powers are lowered by one. Within certain restrictions, a Taylor series expansion of the location vectors and rotation tensors up to the first term is sometimes sufficient, yet this is not the general case. This fact is not taken into consideration in many papers, and it is occasionally overlooked in the creation of multibody dynamics software.

The linearization introduced here assumes continuity in the nonlinear relations, which is indeed generally the case in kinematics. On the other hand, it is possible for the applied forces to have an unsteady development, such as in Coulomb friction. Then complete linearization is impossible, and other methods must be employed, e.g. harmonic linearization. The reader is referred to the relevant literature for further details, see e.g. Nayfeh and Mook [38] or Sextro, Popp, and Magnus [53].

Example 5.3 (Physical Pendulum). The equation of motion of a physical pendulum, see Fig. 5.4, will be linearized with respect to the zero position $\alpha_S(t) \equiv 0$, i.e.,

$$\alpha = \alpha_S + \alpha_L = \alpha_L. \tag{5.63}$$

First, the linear differential equation of a conservative oscillator follows directly from the nonlinear equation of motion with $\sin\alpha_L \approx \alpha_L$,

$$5mL^2\ddot{\alpha}_L(t) = -3mgL\alpha_L(t). \tag{5.64}$$

Now we will show how to linearize the kinematics. For the location vector and the rotation tensor, we obtain with $\sin\alpha_L \approx \alpha_L$ and $\cos\alpha_L \approx 1 - \frac{1}{2}\alpha_L^2$ the kinematics relations up to the second term

$$\boldsymbol{r}(\alpha_L) = \begin{bmatrix} 0 \\ \frac{3}{2}L\alpha_L \\ -\frac{3}{2}L(1-\frac{1}{2}\alpha_L^2) \end{bmatrix}, \tag{5.65}$$

$$\boldsymbol{S}(\alpha_L) = \begin{bmatrix} 1 & 0 & 0 \\ 0 & 1-\frac{1}{2}\alpha_L^2 & -\alpha_L \\ 0 & \alpha_L & 1-\frac{1}{2}\alpha_L^2 \end{bmatrix}. \tag{5.66}$$

The linearized global Jacobian matrix is then

$$\overline{\boldsymbol{J}}^T = \begin{bmatrix} 0 & \frac{3}{2}L & \frac{3}{2}L\alpha_L & 1 & 0 & 0 \end{bmatrix}. \tag{5.67}$$

The Newton-Euler equations take on the form

$$\begin{bmatrix} 0 \\ 3mL \\ 0 \\ \frac{1}{2}mL^2 \\ 0 \\ 0 \end{bmatrix} \ddot{\alpha}_L(t) = \begin{bmatrix} 0 \\ 0 \\ -2mg \\ 0 \\ 0 \\ 0 \end{bmatrix} + \overline{\boldsymbol{q}}^r. \tag{5.68}$$

If we apply (5.67) to (5.68), we immediately obtain (5.64) again due to the orthogonality between reaction forces and motions.

This simplest of all examples already gives us interesting insights. The restoring torque (stiffness) of the physical pendulum is based on the quadratic term in the location vector. If we stop the series expansion after the linear term, the stiffness of the physical pendulum would be lost! In contrast, the quadratic terms originating from the rotation tensor are dropped, the Jacobian matrix of rotation is independent of α_L. It would have sufficed in this special case, therefore, to have expanded the rotation tensor only to the linear term.

End of Example 5.3.

5.3.2 General Multibody Systems

The equations of motion of a nonholonomic multibody system are obtained from (5.19) under consideration of the Jourdain principle (4.25) as

$$\boldsymbol{M}(\boldsymbol{y},\boldsymbol{z},t)\cdot\dot{\boldsymbol{z}}(t)+\boldsymbol{k}(\boldsymbol{y},\boldsymbol{z},t)=\boldsymbol{q}(\boldsymbol{y},\boldsymbol{z},t), \tag{5.69}$$

where the reaction forces again are dropped. In addition, the problem is again symmetrized. The symmetric $g \times g$ inertia matrix reads as

$$\boldsymbol{M}(\boldsymbol{y},\boldsymbol{z},t)=\boldsymbol{L}^T\cdot\overline{\overline{\boldsymbol{M}}}\cdot\boldsymbol{L} \tag{5.70}$$

and again it depends in the rheonomic case explicitly on time. Also appearing are the $g \times 1$ vectors $\boldsymbol{k}$ and $\boldsymbol{q}$ of the generalized gyroscopic forces and the generalized applied forces. There is a close relationship as well between the equations of motion (5.35) for holonomic systems and (5.69). This means that we can also obtain (5.69) from (5.35). For this purpose, we take into account the $f \times g$ functional matrix $\boldsymbol{K}(\boldsymbol{y},\boldsymbol{z},t)$ defined in (2.235). In accordance with (5.38)–(5.40),

$$\boldsymbol{M}(\boldsymbol{y},\boldsymbol{z},t)=\boldsymbol{K}^T\cdot\boldsymbol{M}(\boldsymbol{y},t)\cdot\boldsymbol{K}, \tag{5.71}$$

$$\boldsymbol{k}(\boldsymbol{y},\boldsymbol{z},t)=\boldsymbol{K}^T\cdot\left[\boldsymbol{M}(\boldsymbol{y},t)\cdot\left(\frac{\partial\dot{\boldsymbol{y}}}{\partial\boldsymbol{y}}\cdot\dot{\boldsymbol{y}}+\frac{\partial\dot{\boldsymbol{y}}}{\partial t}\right)+\boldsymbol{k}(\boldsymbol{y},\dot{\boldsymbol{y}},t)\right], \tag{5.72}$$

$$\boldsymbol{q}(\boldsymbol{y},\boldsymbol{z},t)=\boldsymbol{K}^T\cdot\boldsymbol{q}(\boldsymbol{y},\dot{\boldsymbol{y}},t), \tag{5.73}$$

where (2.220) should be inserted each time on the right-hand side.

The equations of motion (5.69) are not solvable alone. They have to be supplemented by the differential equations (2.220) of the nonholonomic constraints. Now proportional-integral forces are also permitted according to (3.12). This means that the solution requires that we investigate the following coupled system of differential equations,

$$\dot{\boldsymbol{y}}=\dot{\boldsymbol{y}}(\boldsymbol{y},\boldsymbol{z},t), \tag{5.74}$$

$$\boldsymbol{M}(\boldsymbol{y},\boldsymbol{z},t)\cdot\dot{\boldsymbol{z}}(t)+\boldsymbol{k}(\boldsymbol{y},\boldsymbol{z},t)=\boldsymbol{q}(\boldsymbol{y},\boldsymbol{z},\boldsymbol{w},t), \tag{5.75}$$

$$\dot{\boldsymbol{w}}=\dot{\boldsymbol{w}}(\boldsymbol{y},\boldsymbol{z},\boldsymbol{w},t). \tag{5.76}$$

This completes the description of general multibody systems.

Example 5.4 (Transport Cart). The transport cart in Figs. 2.18 and 5.5 consists of a rigid body K, which is constrained by massless wheels on a rough inclined plane, and a frictionless caster. The center of mass C should coincide with the axial center P. Then all the kinematic relations given in Example 2.14 for this nonholonomic system are applicable.

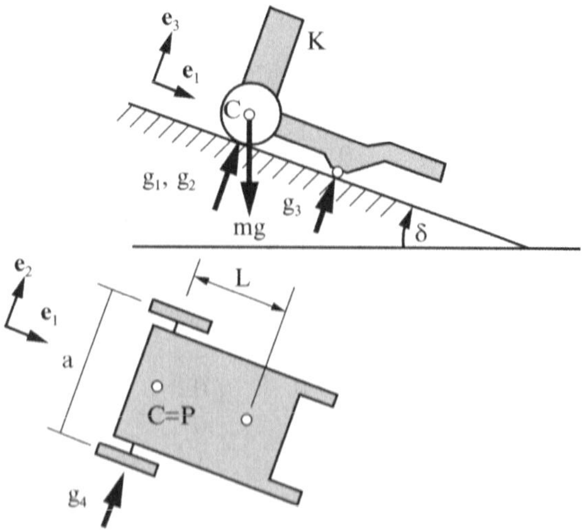

Fig. 5.5 Transport cart on an inclined plane

The mass of the transport cart is m, and the inertia tensor is written in the inertial frame

$$\boldsymbol{I} = \begin{bmatrix} I_{11} & I_{12} & I_{31} \\ I_{12} & I_{22} & I_{23} \\ I_{31} & I_{23} & I_{33} \end{bmatrix}. \tag{5.77}$$

As a result of the four constraints, four generalized reaction forces exist, the normal forces g_1, g_2, and g_3 at both wheels and the caster as well as the cornering force g_4 of the axis. With the distances a, L of the wheels and skid and of the rotation γ around the $\boldsymbol{e}_3$ axis, the reaction forces and torques are written in the inertial frame

$$\boldsymbol{f}^r = \begin{bmatrix} 0 & 0 & 0 & -\sin\gamma \\ 0 & 0 & 0 & \cos\gamma \\ 1 & 1 & 1 & 0 \end{bmatrix} \cdot \begin{bmatrix} g_1 \\ g_2 \\ g_3 \\ g_4 \end{bmatrix}, \tag{5.78}$$

$$\boldsymbol{l}^r = \begin{bmatrix} \frac{a}{2}\cos\gamma & -\frac{a}{2}\cos\gamma & L\sin\gamma & 0 \\ \frac{a}{2}\sin\gamma & -\frac{a}{2}\sin\gamma & -L\cos\gamma & 0 \\ 0 & 0 & 0 & 0 \end{bmatrix} \cdot \begin{bmatrix} g_1 \\ g_2 \\ g_3 \\ g_4 \end{bmatrix}. \tag{5.79}$$

The only applied force is the weight

$$\boldsymbol{f}^e = \begin{bmatrix} mg\sin\delta & 0 & -mg\cos\delta \end{bmatrix}. \tag{5.80}$$

Now the Newton-Euler equations read

$$\underbrace{\begin{bmatrix} m\cos\gamma & 0 \\ m\sin\gamma & 0 \\ 0 & 0 \\ 0 & -I_{31} \\ 0 & -I_{23} \\ 0 & I_{33} \end{bmatrix}}_{\overline{\overline{\boldsymbol{M}}}\cdot\overline{\boldsymbol{L}}(\boldsymbol{y})} \cdot \underbrace{\begin{bmatrix} \dot{v} \\ \ddot{\gamma} \end{bmatrix}}_{\dot{\boldsymbol{z}}(t)} + \underbrace{\begin{bmatrix} -mv\dot{\gamma}\sin\gamma \\ mv\dot{\gamma}\cos\gamma \\ 0 \\ I_{23}\dot{\gamma}^2 \\ -I_{31}\dot{\gamma}^2 \\ 0 \end{bmatrix}}_{\overline{\boldsymbol{q}}^c(\boldsymbol{y},\boldsymbol{z})} = \underbrace{\begin{bmatrix} mg\sin\delta \\ 0 \\ -mg\cos\delta \\ 0 \\ 0 \\ 0 \end{bmatrix}}_{\overline{\boldsymbol{q}}^e}$$

$$+ \underbrace{\begin{bmatrix} 0 & 0 & 0 & -\sin\gamma \\ 0 & 0 & 0 & \cos\gamma \\ 1 & 1 & 1 & 0 \\ \frac{a}{2}\cos\gamma & -\frac{a}{2}\cos\gamma & L\sin\gamma & 0 \\ \frac{a}{2}\sin\gamma & -\frac{a}{2}\sin\gamma & -L\cos\gamma & 0 \\ 0 & 0 & 0 & 0 \end{bmatrix}}_{\overline{\boldsymbol{Q}}(\boldsymbol{y})} \cdot \underbrace{\begin{bmatrix} g_1 \\ g_2 \\ g_3 \\ g_4 \end{bmatrix}}_{\boldsymbol{g}(t)} . \tag{5.81}$$

After left-multiplication with the transposed global Jacobian matrix $\overline{\boldsymbol{L}}^T$, we obtain directly the equations of motion

$$\underbrace{\begin{bmatrix} m & 0 \\ 0 & I_{33} \end{bmatrix}}_{\boldsymbol{M}} \cdot \underbrace{\begin{bmatrix} \dot{v} \\ \ddot{\gamma} \end{bmatrix}}_{\dot{\boldsymbol{z}}(t)} = \underbrace{\begin{bmatrix} mg\sin\delta\cos\gamma \\ 0 \end{bmatrix}}_{\boldsymbol{q}(\boldsymbol{y})} . \tag{5.82}$$

In conjunction with the kinematic equations (2.239), the equations of motion (5.82) completely describe the given general multibody system. Both differential equations can be solved in a closed form in this simple example. The second differential equation of (5.82) with the initial conditions $\gamma_0 = 0$, $\dot{\gamma}_0 = \Omega$ yields a constant rotational velocity

$$\gamma(t) = \Omega t. \tag{5.83}$$

With this we obtain from the first differential equation of (5.82) with the initial condition $v_0 = 0$ the periodically variable velocity

$$v(t) = \frac{1}{\Omega} g \sin\delta \sin\Omega t. \tag{5.84}$$

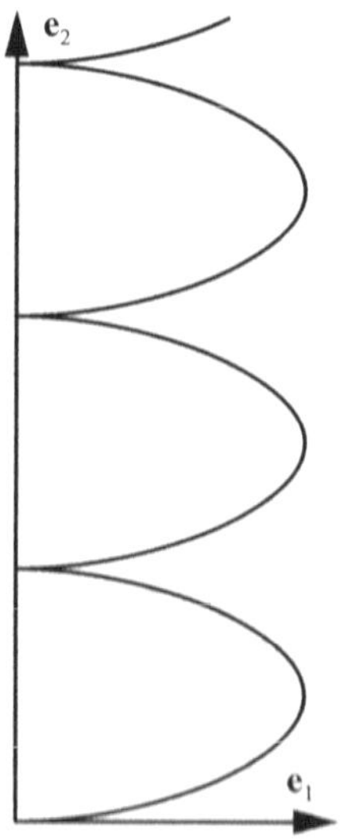

Fig. 5.6 Path of the center of mass C of the transport cart

However, the equations of motion do not tell us anything about the path of the center of mass. This can only be found from the kinematics of the nonholonomic constraints (2.239), which yield further differential equations

$$\dot{r}_1 = \frac{g \sin \delta}{\Omega} \sin \Omega t \cos \Omega t, \tag{5.85}$$

$$\dot{r}_2 = \frac{g \sin \delta}{\Omega} \sin^2 \Omega t. \tag{5.86}$$

With the initial conditions $r_{10} = r_{20} = 0$, we find after another time integration

$$r_1(t) = \frac{1}{4} \frac{g \sin \delta}{\Omega^2} (1 - \cos 2\Omega t), \tag{5.87}$$

$$r_2(t) = \frac{1}{4} \frac{g \sin \delta}{\Omega^2} (2\Omega t - \sin 2\Omega t). \tag{5.88}$$

The transport cart thus moves cross to the plane in the 2-direction and does not roll down the plane in 1-direction, see Fig. 5.6. Only for $\Omega = 0$ do we obtain according to l'Hospital's rule

$$r_1(t) = \frac{1}{2} g t^2 \sin \delta, \qquad r_2(t) = 0, \tag{5.89}$$

i.e. the transport cart then behaves like a mass point.

End of Example 5.4.

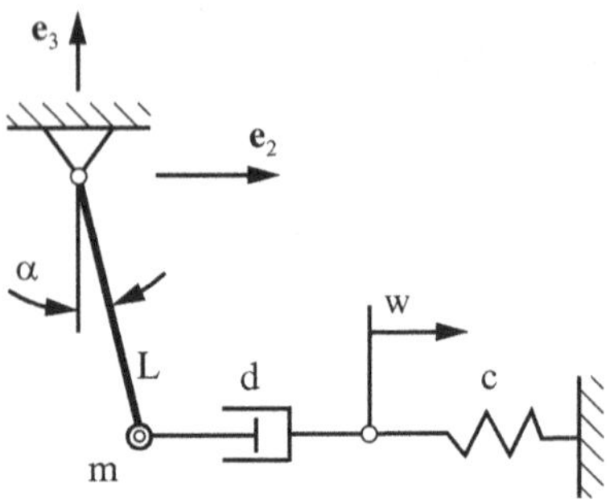

Fig. 5.7 Pendulum with an elastic damper

Example 5.5 (Elastic Damper). The small oscillations of a mathematical pendulum should be damped. To avoid hard impacts, an elastic damper is used, see Fig. 5.7. This results in a system with proportional-integral forces. For small angles, i.e. for $\alpha \ll 1$, we find the Newton equations

$$m\begin{bmatrix} 0 \\ L \\ L\alpha \end{bmatrix} \ddot{\alpha}(t) = \begin{bmatrix} 0 \\ -cw(t) \\ -mg \end{bmatrix} + \overline{\boldsymbol{q}}^r, \tag{5.90}$$

where we dispense with explicitly determining the reaction forces, since these are eliminated anyhow. After left-multiplication with the transposed Jacobian matrix we obtain

$$mL^2\ddot{\alpha}(t) + mgL\alpha(t) + Lcw(t) = 0. \tag{5.91}$$

This equation of motion cannot be solved itself, since $w(t)$ is unknown. The equation of motion (5.91) must be supplemented by the differential equation of the force variable $w(t)$. From the reaction law for the cutting forces

$$d(L\dot{\alpha}(t) - \dot{w}(t)) = cw(t) \tag{5.92}$$

it follows that

$$\dot{w}(t) = L\dot{\alpha}(t) - \frac{c}{d}w(t). \tag{5.93}$$

Equations (5.91) and (5.93) describe a general multibody system.

End of Example 5.5.

A general multibody system of the form (5.74) cannot be transformed into the form (5.39) of an ordinary multibody system. However, an ordinary multibody system can be represented the other way round in the form (5.74), which entails a separation of kinematics and kinetics. To this end, we introduce in addition to the $f \times 1$ position vector $\mathbf{y}(t)$ of the generalized coordinates a second $f \times 1$ vector $\mathbf{z}(t)$, which

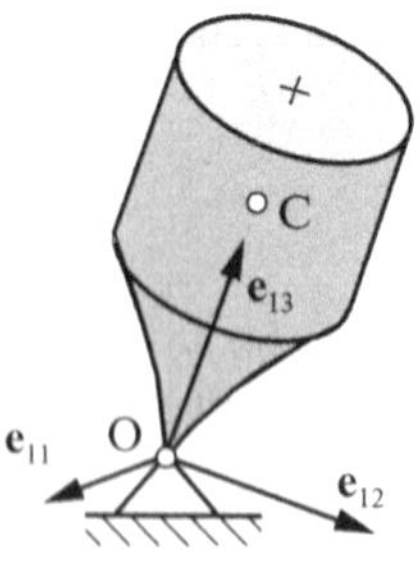

Fig. 5.8 Heavy gyroscope

describes the generalized velocities. Both vectors should be linked by a regular $f \times f$ matrix $\boldsymbol{K}(\mathbf{y},t)$,

$$\dot{\mathbf{y}}(t) = \boldsymbol{K}(\mathbf{y},t) \cdot \mathbf{z}(t), \tag{5.94}$$

which corresponds formally to a nonholonomic constraint according to (2.220) and (2.235). However, since $f = g$ applies, it also follows from (2.218) that (5.94) does not represent a real constraint, since no degrees of freedom have been lost, $r = 0$. After left-multiplication with $\boldsymbol{K}^T$, (5.35) and (5.94) yield the equations of motion

$$\begin{aligned} &\boldsymbol{K}^T(\mathbf{y},t) \cdot \boldsymbol{M}(\mathbf{y},t) \cdot \boldsymbol{K}(\mathbf{y},t) \cdot \dot{\mathbf{z}}(t) \\ &+\boldsymbol{K}^T(\mathbf{y},t) \cdot \left[\boldsymbol{M}(\mathbf{y},t) \cdot \dot{\boldsymbol{K}}(\mathbf{y},\dot{\mathbf{y}},t) \cdot \mathbf{z}(t) + \boldsymbol{k}(\mathbf{y},\dot{\mathbf{y}},t)\right] = \boldsymbol{K}^T(\mathbf{y},t) \cdot \boldsymbol{q}(\mathbf{y},\dot{\mathbf{y}},t), \end{aligned} \tag{5.95}$$

which can of course only be solved together with (5.94). Yet the equations of motion (5.95) frequently have a simpler structure than (5.35), a fact that is exploited again and again in gyroscope theory and for large, nonlinear multibody systems.

Example 5.6 (Heavy Gyroscope). A gyroscope is a rigid body with a fixed point, see Fig. 5.8. In this case, we can merge the Newton and Euler equations (3.22) and (3.30) for the center of mass C into new Euler equations, which then however apply to the fixed point O as a reference point. This property is interesting for a single rigid body, and gyroscope theory makes vivid use of it. Moreover, a body-fixed frame is always used as a reference frame R.

If we chose the Euler angles as the generalized coordinates

$$\mathbf{y}(t) = \begin{bmatrix} \psi & \vartheta & \phi \end{bmatrix} \tag{5.96}$$

with the rotation tensor (2.58), then it follows for the rotational velocity vector in compliance with (2.100),

$$\boldsymbol{\omega} = \boldsymbol{J}_R(\mathbf{y}) \cdot \dot{\mathbf{y}}(t) \tag{5.97}$$

and in accordance with (2.121) for the rotational acceleration vector

$$\boldsymbol{\alpha} = \boldsymbol{J}_R(\boldsymbol{y}) \cdot \ddot{\boldsymbol{y}}(t) + \left(\frac{\partial \boldsymbol{J}_R(\boldsymbol{y})}{\partial \boldsymbol{y}} \cdot \dot{\boldsymbol{y}}(t) \right) \cdot \dot{\boldsymbol{y}}(t) \tag{5.98}$$

in the body-fixed frame. The equations of motion are written according to the d'Alembert principle in the body fixed frame

$$\boldsymbol{J}_R^T(\boldsymbol{y}) \cdot \boldsymbol{I} \cdot \boldsymbol{J}_R(\boldsymbol{y}) \cdot \ddot{\boldsymbol{y}}(t) + \boldsymbol{k}(\boldsymbol{y},\dot{\boldsymbol{y}}) = \boldsymbol{q}(\boldsymbol{y}), \tag{5.99}$$

where the torque of the weight has been taken into account. The inertia matrix is highly nonlinear, so an analytical solution appears not to be possible.

If, in addition to the generalized coordinates, we introduce the rotational velocities

$$\boldsymbol{z}(t) = \left[\begin{array}{ccc} \omega_1 & \omega_2 & \omega_3 \end{array} \right] \tag{5.100}$$

as generalized velocities, then

$$\boldsymbol{\omega} = \boldsymbol{L}_R \cdot \boldsymbol{z}(t), \qquad \boldsymbol{\alpha} = \boldsymbol{L}_R \cdot \dot{\boldsymbol{z}}(t) \tag{5.101}$$

with $\boldsymbol{L}_R = \boldsymbol{E} =$ constant. This simplifies the equations of motion considerably

$$\boldsymbol{I} \cdot \dot{\boldsymbol{z}}(t) + \boldsymbol{k}(\boldsymbol{z}) = \boldsymbol{q}(\boldsymbol{y}), \tag{5.102}$$

i.e. the time-invariant inertia tensor in the body-fixed frame with respect to the fixed point O remains as inertia matrix. But we can also see that the equations of motion (5.102) cannot be solved alone. They have to be supplemented by (5.94) with $\boldsymbol{K}(\boldsymbol{y}) = \boldsymbol{J}_R^{-1}(\boldsymbol{y})$ following by comparison with (5.97). Nonetheless, the heavy pendulum is an ordinary multibody system because the transformation of (5.102) with (5.71) and (5.72) to the form (5.99) is possible.

It should also be noted that (5.102) and (5.94) can be completely solved analytically in the case of symmetry $I_{11} = I_{22}$. The solution traces back to Lagrange and is extensively described in the literature, see Magnus [35] or Arnold and Maunder [2].

End of Example 5.6.

The transformation (5.94) can also be used to represent the equations of motion of ordinary multibody systems in a normal form. For this purpose, we should take into account that each positive definite matrix can be converted into a unit matrix by a congruence transformation, which corresponds to a sequence of elementary matrix operations.

The normal form of the equations of motion (5.35) is given by (5.94) and

$$\dot{\boldsymbol{z}}(t) = \boldsymbol{K}^T(\boldsymbol{y},t) \cdot \left[\boldsymbol{q} - \boldsymbol{k} - \boldsymbol{M} \cdot \dot{\boldsymbol{K}}(\boldsymbol{y},t) \cdot \boldsymbol{z} \right], \tag{5.103}$$

where the arguments of the values in the brackets are not written. The transformation matrix $\boldsymbol{K}(\mathbf{y},t)$ must satisfy the conditions

$$\boldsymbol{K}^T(\mathbf{y},t)\cdot\boldsymbol{M}(\mathbf{y},t)\cdot\boldsymbol{K}(\mathbf{y},t)=\boldsymbol{E}, \tag{5.104}$$

which converts the $f\times f$ inertia matrix into the $f\times f$ unit matrix $\boldsymbol{E}$. The normal form (5.94), (5.103) can sometimes have numerical advantages, since the inversion of the inertia matrix is omitted in the integration.

Example 5.7 (Normal Form of an Inertia Matrix). The inertia matrix of the double pendulum is given by (5.45),

$$\boldsymbol{M}=mL^2\begin{bmatrix} 2 & \cos(\alpha_1-\alpha_2) \\ \cos(\alpha_1-\alpha_2) & 1 \end{bmatrix}. \tag{5.105}$$

First the matrix (5.105) is diagonalized by removing the second column/line from the first column/line after multiplication with $\cos(\alpha_1-\alpha_2)$. Then the first column and line are multiplied by the reciprocal of the root of the first diagonal element. These elementary operations lead to the transformation matrix

$$\boldsymbol{K}=\frac{1}{\sqrt{mL^2}}\begin{bmatrix} \dfrac{1}{\sqrt{1+\sin^2(\alpha_1-\alpha_2)}} & 0 \\ \dfrac{-\cos(\alpha_1-\alpha_2)}{\sqrt{1+\sin^2(\alpha_1-\alpha_2)}} & 1 \end{bmatrix}. \tag{5.106}$$

We can finally see that (5.105) and (5.106) satisfy the relation (5.104).

End of Example 5.7.

5.4 Equations of Reaction of Ideal Systems

When setting up equations of motion, the reaction forces are omitted. Yet in order to design the constraint elements (joints, bearings) and to estimate the strength of machine components, the reaction forces are of great importance. The external reaction forces also determine a machine's impact on the environment, which can however be reduced or eliminated by a suitable internal mass balance.

5.4.1 *Calculating Reaction Forces*

In order to calculate reaction forces, we have to return to the Newton-Euler equations (5.18). Direct evaluation of these equations is not favorable, however. First of all, the generalized accelerations $\ddot{\mathbf{y}}(t)$ are annoying, and secondly (5.18) is overdetermined by the $6p\times q$ distribution matrix $\overline{\boldsymbol{Q}}$ with respect to the vector of the generalized reaction forces $\boldsymbol{g}(t)$. Both problems can be solved with the orthogonality condition (4.16) corresponding to the principle of virtual work, see e.g. Schiehlen [50]. If we multiply (5.18) from the left with the $q \times 6p$ matrix $\overline{\boldsymbol{Q}}^T \cdot \overline{\overline{\boldsymbol{M}}}^{-1}$, we then obtain the equations of reaction in the form of a linear algebraic system of equations

$$\boldsymbol{N}(\boldsymbol{y},t) \cdot \boldsymbol{g}(t) = \hat{\boldsymbol{k}}(\boldsymbol{y},\dot{\boldsymbol{y}},t) - \hat{\boldsymbol{q}}(\boldsymbol{y},\dot{\boldsymbol{y}},t). \tag{5.107}$$

Here,

$$\boldsymbol{N}(\boldsymbol{y},t) = \overline{\boldsymbol{Q}}^T \cdot \overline{\overline{\boldsymbol{M}}}^{-1} \cdot \overline{\boldsymbol{Q}} \tag{5.108}$$

is a symmetric, generally positive definite $q \times q$ reaction matrix, while the $q \times 1$ vectors $\hat{\boldsymbol{q}}(\boldsymbol{y},\dot{\boldsymbol{y}},t)$ and $\hat{\boldsymbol{k}}(\boldsymbol{y},\dot{\boldsymbol{y}},t)$ represent the influence of applied forces and gyroscopic forces on the reaction forces.

When setting up equations of motion, we can completely dispense with determining the reaction forces and in particular with determining the distribution matrix $\overline{\boldsymbol{Q}}$, since the motion is not affected by it. In order to establish equations of reaction on the other hand, the distribution matrix $\overline{\boldsymbol{Q}}$ is always necessary. A few points will be made on this subject. The first possibility is the descriptive-constructive method. In this case, to each constraint in a frame oriented at the joint axes a corresponding reaction force is assigned. By means of transformation, the reaction forces are then converted into the global frame, the inertial frame for example. The second way to determine the distribution matrix is given by (4.13), i.e. using implicit constraints. In this method, implicit constraints can be either directly derived using the Cartesian coordinates of the inertial frame or the derivation is carried out using generalized coordinates for the free system with subsequent consideration of the functional matrices (4.13), respectively.

In order to support our intuition of this, according to (4.11) the q generalized reaction forces g_k can be assigned individually to the reaction forces and torques

$$\boldsymbol{f}_{ik}^r = \frac{\partial \phi_k}{\partial \boldsymbol{x}} \cdot \frac{\partial \boldsymbol{x}}{\partial \boldsymbol{r}_i} g_k, \qquad \boldsymbol{l}_{ik}^r = \frac{\partial \phi_k}{\partial \boldsymbol{x}} \cdot \frac{\partial \boldsymbol{x}}{\partial \boldsymbol{s}_i} g_k, \qquad i = 1(1)p,\ k = 1(1)q. \tag{5.109}$$

In (5.109), the 3×1 vector $\boldsymbol{s}_i$ of infinitesimal rotation appears. The relations (5.109) are always useful if there is a lack of clarity concerning the amount, direction, or sense of direction of the reactions to a certain constraint ϕ_k.

If the equations of motion (5.35) with (5.40) are obtained from the equations of motion (5.28), then these relations apply for the equations of reaction (5.107),

$$\boldsymbol{N}_{\text{holonomic}}(\boldsymbol{y},t) = \overline{\boldsymbol{G}}^T \cdot \boldsymbol{M}_{\text{free}}^{-1}(\boldsymbol{x},t) \cdot \overline{\boldsymbol{G}}, \tag{5.110}$$

$$\hat{\boldsymbol{k}}_{\text{holonomic}}(\boldsymbol{y},\dot{\boldsymbol{y}},t) = \overline{\boldsymbol{G}}^T \cdot [\dot{\boldsymbol{I}} \cdot \dot{\boldsymbol{y}} + \boldsymbol{M}_{\text{free}}^{-1} \cdot \boldsymbol{k}_{\text{free}}(\boldsymbol{x},\dot{\boldsymbol{x}})], \tag{5.111}$$

$$\hat{\boldsymbol{q}}_{\text{holonomic}}(\boldsymbol{y},\dot{\boldsymbol{y}},t) = \overline{\boldsymbol{G}}^T \cdot \boldsymbol{M}_{\text{free}}^{-1} \cdot \boldsymbol{q}_{\text{free}}(\boldsymbol{x},\dot{\boldsymbol{x}},t). \tag{5.112}$$

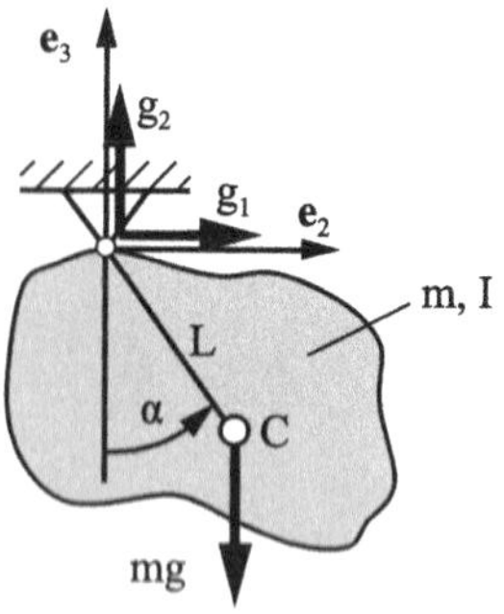

Fig. 5.9 Reaction forces acting on a physical pendulum

The arguments on the right-hand side must again be substituted with (5.40). Comparison of (5.108) with (5.110) shows us the following: On the one hand, inversion of the block diagonal matrix $\overline{\overline{\boldsymbol{M}}}$ is simpler than inversion of $\boldsymbol{M} = \overline{\boldsymbol{H}}^T \cdot \overline{\overline{\boldsymbol{M}}} \cdot \overline{\boldsymbol{H}}$, on the other hand the matrix $\overline{\boldsymbol{G}}$ defined in accordance with (4.13) is generally less fully occupied than $\overline{\boldsymbol{Q}} = \overline{\boldsymbol{H}}^{-T} \cdot \overline{\boldsymbol{G}}$. It is thus recommendable to compute the distribution matrix $\overline{\boldsymbol{Q}}$ first from $\overline{\boldsymbol{G}}$ and $\overline{\boldsymbol{H}}$ and then to return to the Newton-Euler equations according to (5.108).

Example 5.8 (Planar Physical Pendulum). For the physical pendulum, see Fig. 5.9, the implicit constraints are

$$\boldsymbol{\phi} = \begin{bmatrix} r_2 - L\sin\alpha \\ r_3 + L\cos\alpha \end{bmatrix} = \mathbf{0} \tag{5.113}$$

with the three coordinates of the planar motion, which can be summarized in the vector

$$\boldsymbol{x} = \begin{bmatrix} r_2 & r_3 & \alpha \end{bmatrix}. \tag{5.114}$$

The global 3×3 Jacobian matrix of the free system is thereby equal to the unit matrix, $\overline{\boldsymbol{H}} = \boldsymbol{E}$. The 3×2 distribution matrix reads as

$$\overline{\boldsymbol{Q}} = \overline{\boldsymbol{G}} = \frac{\partial \boldsymbol{\phi}}{\partial \boldsymbol{x}} = \begin{bmatrix} 1 & 0 & -L\cos\alpha \\ 0 & 1 & -L\sin\alpha \end{bmatrix}. \tag{5.115}$$

The resultant Newton-Euler equations in the inertial frame are

$$\begin{bmatrix} mL\cos\alpha \\ mL\sin\alpha \\ I \end{bmatrix} \ddot{\alpha} + mL \begin{bmatrix} -\dot{\alpha}^2 \sin\alpha \\ \dot{\alpha}^2 \cos\alpha \\ 0 \end{bmatrix}$$

$$= \begin{bmatrix} 0 \\ -mg \\ 0 \end{bmatrix} + \begin{bmatrix} 1 & 0 \\ 0 & 1 \\ -L\cos\alpha & -L\sin\alpha \end{bmatrix} \cdot \begin{bmatrix} g_1 \\ g_2 \end{bmatrix}. \tag{5.116}$$

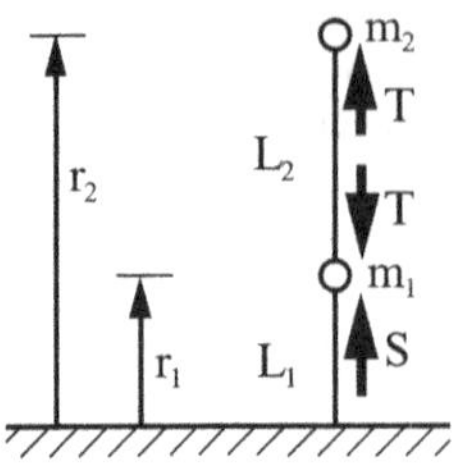

Fig. 5.10 Reaction forces of a static point system

After left-multiplication with $\overline{\boldsymbol{Q}}^T \cdot \overline{\overline{\boldsymbol{M}}}^{-1}$, we have

$$\begin{bmatrix} \frac{1}{m}+\frac{L^2}{I}\cos^2\alpha & \frac{L^2}{I}\sin\alpha\cos\alpha \\ \frac{L^2}{I}\sin\alpha\cos\alpha & \frac{1}{m}+\frac{L^2}{I}\sin^2\alpha \end{bmatrix} \cdot \begin{bmatrix} g_1 \\ g_2 \end{bmatrix} + \begin{bmatrix} 0 \\ -g \end{bmatrix} = \begin{bmatrix} -L\dot{\alpha}^2\sin\alpha \\ L\dot{\alpha}^2\cos\alpha \end{bmatrix}. \quad (5.117)$$

This linear system of equations clearly determines the bearing forces of the physical pendulum. According to (5.109), g_1 corresponds to the horizontal reaction force and g_2 to the vertical reaction force, as is shown in Fig. 5.9.

End of Example 5.8.

Example 5.9 (Transport Cart). The Newton-Euler equations (5.81) contain a distribution matrix $\overline{\boldsymbol{Q}}$, which is compiled descriptively according to Fig. 5.5. It is easy to convince oneself that the generalized reaction forces can be determined by left-multiplication with the matrix $\overline{\boldsymbol{Q}}^T \cdot \overline{\overline{\boldsymbol{M}}}^{-1}$ even in the case of nonholonomic systems. Inversion of the 3×3 inertia tensor (5.77) appearing in $\overline{\overline{\boldsymbol{M}}}$ can be executed symbolically without great difficulties with a formula manipulation program.

The reaction forces can of course also be determined in the static case of $f=0$. Then all accelerations are omitted, eliminating the need for left-multiplication of the Newton-Euler equations. Then the time-invariant equilibrium conditions of statics are valid

$$\overline{\boldsymbol{Q}}\cdot\boldsymbol{g}+\overline{\boldsymbol{q}}^e=\boldsymbol{0}. \quad (5.118)$$

If we obtain the $6p\times 6p$ matrix $\overline{\boldsymbol{Q}}$ from the implicit constraints, we automatically find the cutting forces in the bearings between the individual partial bodies of a multibody system.

End of Example 5.9.

Example 5.10 (Static Point System). The point system outlined in Fig. 5.10 is characterized by the two implicit constraints

$$\phi_1 = r_1 - L_1 = 0, \qquad \phi_2 = r_2 - r_1 - L_2 = 0. \quad (5.119)$$

Fig. 5.11 Planar cut through a rigid body

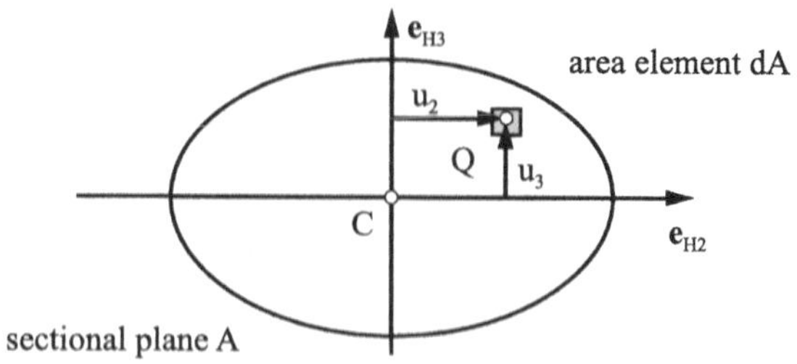

Thus the equilibrium conditions are written with the 2×2 distribution matrix $\overline{\boldsymbol{Q}}$,

$$\begin{bmatrix} 1 & -1 \\ 0 & 1 \end{bmatrix} \cdot \begin{bmatrix} S \\ T \end{bmatrix} + \begin{bmatrix} -m_1 g \\ -m_2 g \end{bmatrix} = \mathbf{0}, \tag{5.120}$$

from which follows the solution

$$\begin{bmatrix} S & T \end{bmatrix} = \begin{bmatrix} (m_1 + m_2)g & m_2 g \end{bmatrix}. \tag{5.121}$$

The generalized constraints S and T are shown in Fig. 5.10.

End of Example 5.10.

5.4.2 *Strength Estimation*

The internal stresses of a body, which are of decisive importance for strength calculation, cannot be determined for rigid bodies because it is a statically indeterminate problem. But it is possible to estimate strength. For this purpose, a linear stress distribution is assumed and calculated from the reaction wrench for an arbitrarily chosen sectional plane, see Fig. 3.6. The resultant maximum stresses can then be used as a basis for the strength estimation. This procedure has long been known in the field of beam statics. It can be transferred to an arbitrary rigid body if we consider the experiences summarized in the principle of De Saint-Venant, namely that individual stress peaks in a continuum have a local effect only. The continuum naturally balances out highly varying stresses.

For an arbitrarily chosen sectional plane A, first the area centroid C and the principal axes frame H are determined in accordance with Fig. 5.11. Then the following applies for the coordinates of the area centroid

$$\int_A u_2 dA = 0, \qquad \int_A u_3 dA = 0 \tag{5.122}$$

and the deviation moment of area is

$$\int_A u_2 u_3 dA = 0. \tag{5.123}$$

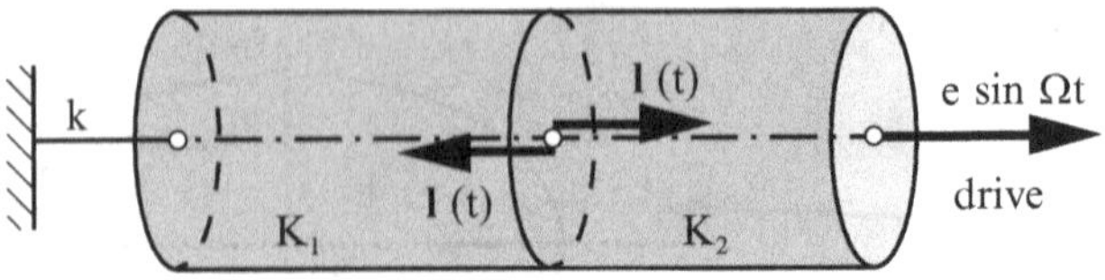

Fig. 5.12 Rigid round bar with harmonic drive torque

The local 3×1 stress vector $\boldsymbol{t}$ at point Q is described by the linear shape function

$$\boldsymbol{t} = \frac{d\boldsymbol{f}}{dA} = \begin{bmatrix} \sigma_{11} \\ \tau_{12} \\ \tau_{13} \end{bmatrix} = \begin{bmatrix} \sigma_1 + \sigma_2 u_3 + \sigma_3 u_2 \\ \tau_2 - \tau_1 u_3 \\ \tau_3 + \tau_1 u_2 \end{bmatrix} \tag{5.124}$$

with the six coefficients $\sigma_1, \sigma_2, \sigma_3, \tau_1, \tau_2, \tau_3$. These can be determined uniquely from the reaction wrench $(\boldsymbol{f}, \boldsymbol{l})$. The reaction wrench can be determined in turn from the equations of reaction (5.107) if we also take the additional rigid body resulting from the cut into consideration in the Newton-Euler equations. Both bodies generated by sectioning are constrained by a firm clamping. Specifically, when integrating over the sectional plane A the following applies for the coordinates of the reaction wrench

$$f_1 = \int_A \sigma_{11} dA, \qquad f_2 = \int_A \tau_{12} dA, \qquad f_3 = \int_A \tau_{13} dA, \tag{5.125}$$

$$l_1 = \int_A (u_2 \tau_{13} - u_3 \tau_{12}) dA, \qquad l_2 = \int_A u_3 \sigma_{11} dA, \qquad l_3 = -\int_A u_2 \sigma_{11} dA. \tag{5.126}$$

If we now insert (5.124) in sequence into (5.125) and (5.126), we find, under consideration of (5.122) and (5.123), the coefficients

$$\sigma_1 = \frac{f_1}{A}, \qquad \tau_2 = \frac{f_2}{A}, \qquad \tau_3 = \frac{f_3}{A}, \tag{5.127}$$

$$\tau_1 = \frac{l_1}{J_P}, \qquad \sigma_2 = \frac{l_2}{J_2}, \qquad \sigma_3 = -\frac{l_3}{J_3}. \tag{5.128}$$

Here, A is the area of the sectional plane, J_2 and J_3 are the axial inertia moments of area, and $J_P = J_2 + J_3$ is the polar inertia moment of area, each with respect to the area centroid. In this way, we can estimate the stresses at any point of the sectional plane.

Example 5.11 (Round Bar). A rigid round bar (mass $2m$, radius R) is fixed on the left by a torsional spring (spring constant k) and is loaded on the right by a harmonic torque (amplitude e, frequency Ω), see Fig. 5.12. The stresses in the center of the bar have to be estimated.

For the purpose of stress estimation, the round bar is cut in the middle, creating two bodies K_1 and K_2. The free rotation motion with respect to the 1-axis is denoted by two generalized coordinates

$$\boldsymbol{x}(t) = \begin{bmatrix} \alpha_1 & \alpha_2 \end{bmatrix}, \tag{5.129}$$

the rigid constraint between the bodies is written implicitly

$$\phi = \alpha_1 - \alpha_2 = 0 \tag{5.130}$$

or explicitly with the scalar generalized position coordinate α

$$\boldsymbol{x} = \begin{bmatrix} 1 & 1 \end{bmatrix} \alpha. \tag{5.131}$$

The Euler equations result in

$$\frac{1}{2} mR^2 \ddot{\alpha}(t) = -k\alpha(t) + l(t), \tag{5.132}$$

$$\frac{1}{2} mR^2 \ddot{\alpha}(t) = e \sin \Omega t - l(t), \tag{5.133}$$

from which result the equation of motion

$$mR^2 \ddot{\alpha}(t) + k\alpha(t) = e \sin \Omega t \tag{5.134}$$

and the equation of reaction

$$2l(t) - k\alpha(t) - e \sin \Omega t = 0. \tag{5.135}$$

With the particular solution of the equation of motion

$$\alpha(t) = \frac{1}{k - mR^2\Omega^2} e \sin \Omega t \tag{5.136}$$

we find the reaction torque

$$l(t) = \frac{2k - mR^2\Omega^2}{k - mR^2\Omega^2} \frac{e}{2} \sin \Omega t, \tag{5.137}$$

from which according to (5.128) and (5.124) the stresses τ_{12}, τ_{13} can be estimated, reaching their highest value on the surface $u_2^2 + u_3^2 = R^2$.

End of Example 5.11.

If, in a holonomic system of p rigid bodies and q constraints, a total of n sections are made through the bodies, we then obtain n additional bodies and $6n$ additional constraints. The number of degrees of freedom is not affected by this,

$$f = 6(p+n) - (q+6n) = 6p - q, \tag{5.138}$$

while the number of constraints increases to $q_n = q + 6n$. Furthermore, for each section the inertia related quantities and applied volume forces must be adjusted.

5.4.3 Mass Balancing in Multibody Systems

External reaction forces impact the environment of a machine. Calculating these forces is therefore of great interest. Since the internal forces and torques within the bodies are omitted in the consideration of multibody systems as an overall system according to the reaction laws (3.7) and (3.40), it is advisable to evaluate the external reaction forces not from the equations of reaction (5.107) but directly from the Newton-Euler equations. The external reaction forces and torques are determined solely by the external applied forces and torques as well as the inertia forces and torques. Frequently, the inertia forces and torques are much larger than the applied forces and torques, especially in the case of high-speed machines. In such cases, we restrict ourselves to the investigation and counter balancing of inertia forces and torques, simply called "mass balancing".

Inertia forces and torques appear only in consideration of the overall system when the total momentum and total angular momentum change. Machines with a constant total momentum and constant angular momentum are considered balanced. The demand for constant total momentum is equivalent to the demand for a time-invariant position of the center of mass C of the overall system.

For further calculation, the reaction laws (3.7) and (3.40) are used. For the reference point O, the following applies for the overall system,

$$\sum_{i,j=1}^{p} \boldsymbol{f}_{ij} = \boldsymbol{0}, \qquad \sum_{i,j=1}^{p} (\boldsymbol{l}_{ij} + \tilde{\boldsymbol{r}}_{0i} \cdot \boldsymbol{f}_{ij}) = \boldsymbol{0}, \tag{5.139}$$

or

$$\boldsymbol{G} \cdot \overline{\boldsymbol{q}}^i = \boldsymbol{0}. \tag{5.140}$$

In accordance with (5.23), $\overline{\boldsymbol{q}}^i$ is the $6p \times 1$ vector of the internal forces and torques and

$$\boldsymbol{G} = \left[\begin{array}{cccc|cccc} \boldsymbol{E} & \boldsymbol{E} & \cdots & \boldsymbol{E} & \boldsymbol{0} & \boldsymbol{0} & \cdots & \boldsymbol{0} \\ \hline \tilde{\boldsymbol{r}}_{01} & \tilde{\boldsymbol{r}}_{02} & \cdots & \tilde{\boldsymbol{r}}_{0p} & \boldsymbol{E} & \boldsymbol{E} & \cdots & \boldsymbol{E} \end{array} \right] \tag{5.141}$$

is a $6 \times 6p$ summation matrix, which contains in particular the 3×3 matrices $\tilde{\boldsymbol{r}}_{0i}$ of the location vectors to the centers of mass C_i. The external reactions are found using the Newton-Euler equations (5.18) by left-multiplication with (5.141). This yields the reaction wrench of the overall system

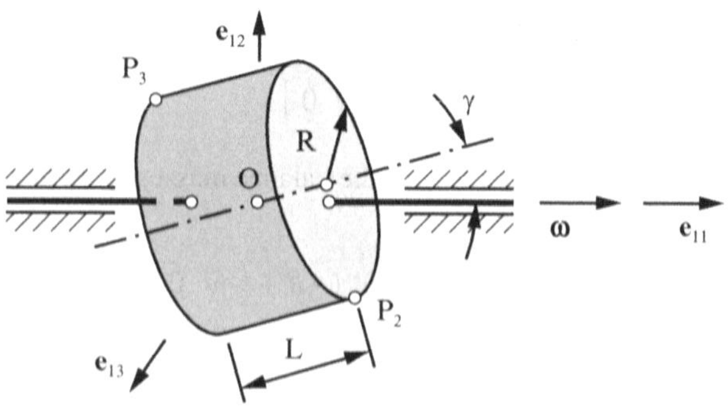

Fig. 5.13 Rigid rotor with balance weights

$$\begin{bmatrix} \boldsymbol{f}_{ra} \\ ---- \\ \boldsymbol{l}_{ra0} \end{bmatrix} = \boldsymbol{G} \cdot (\overline{\overline{\boldsymbol{M}}} \cdot \overline{\boldsymbol{J}} \cdot \ddot{\boldsymbol{y}}(t) + \overline{\boldsymbol{q}}^c - \overline{\boldsymbol{q}}^e). \tag{5.142}$$

If we now neglect the applied forces $\overline{\boldsymbol{q}}^e$, the external reaction forces and torques are equal to the inertia forces and torques, which result from the time derivative of the total momentum and total angular momentum,

$$\boldsymbol{f}_{ra} = \sum_{i=1}^{p} m_i \boldsymbol{a}_i = m\boldsymbol{a}_C, \tag{5.143}$$

$$\boldsymbol{l}_{ra0} = \sum_{i=1}^{p} (\boldsymbol{I}_i \cdot \dot{\boldsymbol{\omega}}_i + \tilde{\boldsymbol{\omega}}_i \cdot \boldsymbol{I}_i \cdot \boldsymbol{\omega}_i + \tilde{\boldsymbol{r}}_{0i} \cdot m_i \boldsymbol{a}_i) \tag{5.144}$$

where $\boldsymbol{a}_C$ is the acceleration of the center of mass C of the overall system. The inertia forces (5.143) are generated by acceleration of the overall center of mass, while the mass moments, according to the three terms in (5.144), can be traced back to a non-uniform rotation, to a dynamic unbalance, or to a static unbalance. For mass balancing, it is now necessary that the inertia forces (5.143) and mass moments (5.144) either vanish or become as small as possible.

Example 5.12 (Balancing a Rigid Rotor). A homogeneous rotor (mass M, length L, radius R) is installed at an inclined angle $\gamma \ll 1$, see Fig. 5.13. To balance the dynamic unbalance, counter balance masses (mass m) are mounted at points P_2 and P_3. The mass required for perfect balancing has to be determined.

In the body-fixed frame, the inertia tensor is

$$\boldsymbol{I} = \begin{bmatrix} I_{11} & I_{12} & 0 \\ I_{12} & I_{22} & 0 \\ 0 & 0 & I_{22} \end{bmatrix}, \qquad I_{12} = -\frac{M}{12}(3R^2 - L^2)\gamma. \tag{5.145}$$

The constant rotational velocity vector is

$$\boldsymbol{\omega} = \begin{bmatrix} \omega & 0 & 0 \end{bmatrix}. \tag{5.146}$$

The location vectors and accelerations of the balance masses P_2, P_3 are obtained e.g. with (2.257)

$$\boldsymbol{r}_{2,3} = \pm \begin{bmatrix} (\frac{L}{2} + R\gamma) & (-R + \frac{L}{2}\gamma) & 0 \end{bmatrix}, \tag{5.147}$$

$$\boldsymbol{a}_{2,3} = \pm \begin{bmatrix} 0 & (-R + \frac{L}{2}\gamma)\omega^2 & 0 \end{bmatrix}. \tag{5.148}$$

Inserted into (5.144), only the 3-coordinate remains for the resultant mass moment in the body-fixed frame as

$$l_{ra03} = \frac{M}{12}(3R^2 - L^2)\omega^2\gamma + mLR\omega^2, \tag{5.149}$$

where $m \ll M$ was additionally taken into account. We can conclude from this that flattened rotors can be balanced with the balance masses at P_2 and P_3. Furthermore, (5.149) also provides the size of the counter balance masses.

End of Example 5.12.

5.5 Equations of Motion and Reaction of Non-ideal Systems

Non-ideal systems are characterized by applied forces also depending on reaction forces. According to (3.13), this is the case e.g. with contact and friction forces. Sliding forces belong to this category, as they depend on the normal pressure and the current direction of the relative velocity. But the cornering forces of an elastic wheel are also applied forces that are determined by the reaction forces. Non-ideal systems thus have definitely real applications.

For non-ideal systems, the equations of motion (5.35) and equations of reaction (5.107) assume the form

$$\boldsymbol{M}(y,t) \cdot \ddot{y}(t) + \boldsymbol{k}(y,\dot{y},t) = \boldsymbol{q}(y,\dot{y},g,t), \tag{5.150}$$

$$\boldsymbol{N}(y,t) \cdot \boldsymbol{g}(t) + \hat{\boldsymbol{q}}(y,\dot{y},g,t) = \hat{\boldsymbol{k}}(y,\dot{y},t). \tag{5.151}$$

We can see that both equations are coupled with each other, and that the equations of reaction (5.151) also can exhibit a nonlinear behavior. This also makes the solution of (5.150) and (5.151) more difficult. Nonetheless, this is usually not crucial. If we solve (5.151) simultaneously during the integration of (5.150), then we always have very good starting values for this nonlinear algebraic system of equations. This is simplified to some extent if the direction of the reaction force is time-invariant, as is the case in the following example.

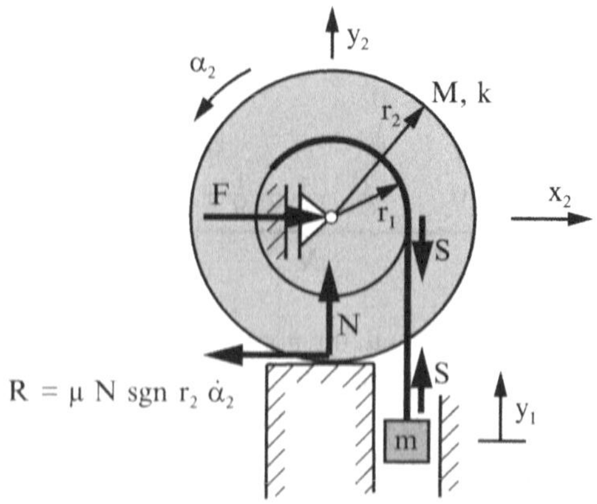

Fig. 5.14 Cable drum with Coulomb friction

Example 5.13 (Cable Drum). The cable drum shown in Fig. 5.14 is sliding on a plane with the sliding friction coefficient μ. With the four essential coordinates of the free system

$$\boldsymbol{x}(t) = \begin{bmatrix} y_1 & x_2 & y_2 & \alpha_2 \end{bmatrix} \tag{5.152}$$

the three constraints are

$$\phi_1 = x_2 = 0, \qquad \phi_2 = y_2 - r_2 = 0, \qquad \phi_3 = y_1 - y_2 - \alpha_2 r_1 = 0, \tag{5.153}$$

leaving $y_1(t)$ as the generalized coordinate. With this we find the explicit constraints

$$\boldsymbol{x} = \begin{bmatrix} y_1 \\ 0 \\ r_2 \\ \dfrac{1}{r_1}(y_1 - y_2) \end{bmatrix}. \tag{5.154}$$

With the parameters plotted in Fig. 5.14, the Newton-Euler equations reads as

$$\begin{bmatrix} m \\ 0 \\ 0 \\ M\dfrac{k^2}{r_1} \end{bmatrix} \ddot{y}_1 = \begin{bmatrix} -mg \\ -\mu N \mathrm{sgn}\dot{y}_1 \\ -Mg \\ -\mu r_2 N \mathrm{sgn}\dot{y}_1 \end{bmatrix} + \begin{bmatrix} 0 & 0 & 1 \\ 1 & 0 & 0 \\ 0 & 1 & -1 \\ 0 & 0 & -r_1 \end{bmatrix} \cdot \begin{bmatrix} F \\ N \\ S \end{bmatrix}. \tag{5.155}$$

From this follows the equation of motion

$$\underbrace{\left(m + M\frac{k^2}{r_1^2}\right)}_{\boldsymbol{M}} \underbrace{\ddot{y}_1}_{\ddot{\boldsymbol{y}}(t)} = \underbrace{-mg - \mu\frac{r_2}{r_1} N \mathrm{sgn}\dot{y}_1}_{\boldsymbol{q}(\dot{\boldsymbol{y}}, g)} \tag{5.156}$$

and the equations of reaction are

$$\underbrace{\begin{bmatrix} \frac{1}{M} & 0 & 0 \\ 0 & \frac{1}{M} & -\frac{1}{M} \\ 0 & -\frac{1}{M} & \frac{1}{m}+\frac{1}{M}+\frac{r_1^2}{Mk^2} \end{bmatrix}}_{N} \cdot \underbrace{\begin{bmatrix} F \\ N \\ S \end{bmatrix}}_{g(t)} + \underbrace{\begin{bmatrix} -\frac{\mu}{M} N \operatorname{sgn} \dot{y}_1 \\ -g \\ \frac{g}{Mk^2} N \operatorname{sgn} \dot{y}_1 \end{bmatrix}}_{\hat{q}(\dot{y},g)} = \mathbf{0}. \qquad (5.157)$$

In the case at hand, the normal force $\boldsymbol{N}$ is time-invariant because $\dot{y}_1(t) < 0$ is always valid given suitable initial conditions. Yet is it obvious that (5.156) cannot be solved without (5.157) even in this case.

End of Example 5.13.

In the investigation of contact problems, elastic multibody systems can also be utilized. As Eberhard [17] has shown, it is often advisable to toggle between models of multibody systems and finite element models in order to enable a more efficient time integration.

5.6 Gyroscopic Equations of Satellites

Multibody systems are also useful for modelling satellites with internally moving masses. The overall motion is then composed of the orbit motion of the center of mass and the rotation motion around the center of mass. Orbit motion is the subject of extensive investigation in celestial mechanics and can be considered as given. The rotational motion, and thus the orientation, obey the laws of gyroscope theory. In particular, the path motion can be eliminated in the gyroscopic equations of the satellite, which means that the number of degrees of freedom is reduced by three.

The housing of the satellite is selected as the base body K_1, see Fig. 5.15. Inside, the bodies $K_j, j = 2(1)p$ are in motion. A housing-fixed frame $\{C_1; \boldsymbol{e}_{1\alpha}\}$, $\alpha = 1(1)3$, in the center of mass C_1 of the housing serves as a moving reference

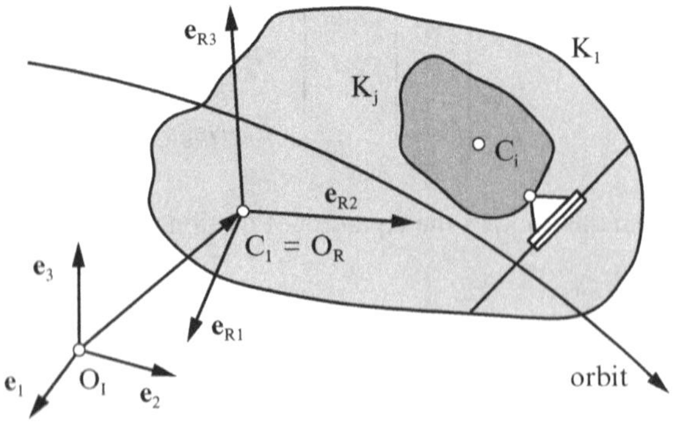

Fig. 5.15 Satellites with internally moving masses

frame $\{0_R; \boldsymbol{e}_{R\alpha}\}$. Then both the $6p \times 1$ position vector $\boldsymbol{x}$ of the free system and the $f \times 1$ position vector $\mathbf{y}$ of the constrained system can be decomposed as follows,

$$\boldsymbol{x}(t) = \begin{bmatrix} \boldsymbol{r}_R \\ -- \\ \boldsymbol{x}_R \end{bmatrix}, \qquad \mathbf{y}(t) = \begin{bmatrix} \boldsymbol{r}_R \\ -- \\ \mathbf{y}_R \end{bmatrix}, \tag{5.158}$$

where $\boldsymbol{r}_R$ represents the 3×1 location vector of the path motion of the housing's center of mass, see Fig. 5.15. With the $3p \times 3$ matrices

$$\overline{\boldsymbol{E}} = \begin{bmatrix} \boldsymbol{E} & \boldsymbol{E} & \dots & \boldsymbol{E} \end{bmatrix}^T, \qquad \overline{\mathbf{0}} = \begin{bmatrix} \mathbf{0} & \mathbf{0} & \dots & \mathbf{0} \end{bmatrix}^T, \tag{5.159}$$

which are built up of 3×3 unit and zero matrices. The global $6p \times f$ Jacobian matrix is

$$\overline{\boldsymbol{J}} = \begin{bmatrix} \overline{\boldsymbol{E}} & | & \overline{\boldsymbol{J}}_{RT} \\ --- & | & ---- \\ \overline{\mathbf{0}} & | & \overline{\boldsymbol{J}}_{RR} \end{bmatrix}. \tag{5.160}$$

Furthermore, if we subdivide the $6p \times 6p$ block diagonal matrix (5.20) into four $3p \times 3p$ matrices

$$\overline{\overline{\boldsymbol{M}}} = \begin{bmatrix} \overline{\overline{\boldsymbol{M}}}_T & | & \mathbf{0} \\ --- & | & --- \\ \mathbf{0} & | & \overline{\overline{\boldsymbol{M}}}_R \end{bmatrix}, \tag{5.161}$$

we then obtain the equations of motion (5.35) in the subdivided form

$$m\ddot{\boldsymbol{r}}_R(t) + \overline{\boldsymbol{E}}^T \cdot \overline{\overline{\boldsymbol{M}}}_T \cdot \overline{\boldsymbol{J}}_{RT} \cdot \ddot{\mathbf{y}}_R(t) + \boldsymbol{k}_r = \boldsymbol{q}_r, \tag{5.162}$$

$$\left[\overline{\boldsymbol{J}}_{RT}^T \cdot \overline{\overline{\boldsymbol{M}}}_T \cdot \overline{\boldsymbol{J}}_{RT} + \overline{\boldsymbol{J}}_{RR}^T \cdot \overline{\overline{\boldsymbol{M}}}_R \cdot \overline{\boldsymbol{J}}_{RR}\right] \cdot \ddot{\mathbf{y}}_R(t) + \overline{\boldsymbol{J}}_{RT}^T \cdot \overline{\overline{\boldsymbol{M}}}_T \cdot \overline{\boldsymbol{E}} \cdot \ddot{\boldsymbol{r}}_R(t) + \boldsymbol{k}_y = \boldsymbol{q}_y \tag{5.163}$$

with the scalar total mass m of the system,

$$m\boldsymbol{E} = \sum_{i=1}^{p} m_i \boldsymbol{E} = \overline{\boldsymbol{E}}^T \cdot \overline{\overline{\boldsymbol{M}}}_T \cdot \overline{\boldsymbol{E}}. \tag{5.164}$$

We see that the path motion $\boldsymbol{r}_R$ in (5.163) can be eliminated by (5.162). The gyroscopic equations for the $(f-3) \times 1$ position vector $\boldsymbol{y}_R$ of the satellite finally assume the form

$$\boldsymbol{M}_R(\boldsymbol{y}_R, t) \cdot \ddot{\boldsymbol{y}}_R(t) + \boldsymbol{k}_R(\boldsymbol{y}_R, \dot{\boldsymbol{y}}_R, t) = \boldsymbol{q}_R. \tag{5.165}$$

However, it must be borne in mind that the generalized forces $\boldsymbol{q}_R$, which include the gravitational moment, can depend on the path motion $\boldsymbol{r}_R$. The path motion $\boldsymbol{r}_R$ of the housing's center of mass can, however, always be set equal to the path motion of the center of mass of the entire satellite because of the small dimensions of a satellite compared to the orbit radius. The path motion of the center of mass is known from the field of celestial mechanics, and in the simplest case it is determined by Kepler's laws.

If the satellite executes large rotations about more than one axis, it is advisable to introduce the rotation velocities as generalized velocities, thus converting (5.165) to the form (5.74) and (5.75). Additional simplifications result if symmetric rotors move with a constant rotation speed inside the satellite. Then it is called a gyrostat, see e.g. Magnus [35], or Wittenburg [65].

5.7 Formalisms for Multibody Systems

The task of setting up equations of motion and reaction is not trivial for large multibody systems, since it involves numerous arithmetic operations. For this reason, computer-based formalisms have been developed since the 1960s which initially provided purely numerical equations, but now also include symbolic ones.

Numerical formalisms provide the coefficients of the system matrices (5.59) in numerical form for linear multibody systems. In nonlinear cases, the numerical formalisms generate the numerical values of the equations of motion (5.35) required for every integration step in the context of a comprehensive simulation program. We obtain as a result trajectories of the generalized coordinates. In contrast, symbolic formalisms provide additional information. In the case of linear multibody systems, we can see what influence the system parameters have on the individual coefficients of the system matrices (5.59). The result is in the same form as that obtained with a traditional paper-and-pencil calculation. A symbolic formalism relieves the engineer of the execution of extensive and often erroneous intermediate calculations, while all relevant information remains available. Even in the nonlinear case, we obtain complete differential equations, which, for example, permit us to recognize the type of couplings in the system. Symbolically determined equations of motion, linear or nonlinear, can then be solved with any available time integration program. The separation of the generation and the solution of equations of motion simplifies the assessment of results considerably.

Furthermore, a distinction is made between non-recursive and recursive formalisms. Recursive formalisms make use of special topological properties of multibody systems in order to increase numerical efficiency.

Commercial programs for the dynamics of multibody systems such as Simpack, Recurdyn, VL Motion, or MSC. Adams allows an efficient analysis of the dynamics with diverse industrial applications. Research codes such as Neweul-M^2 or Robotran offer the additional possibility to deal with both, the modelling approach and the numerical solution method, and to test one's new ideas.

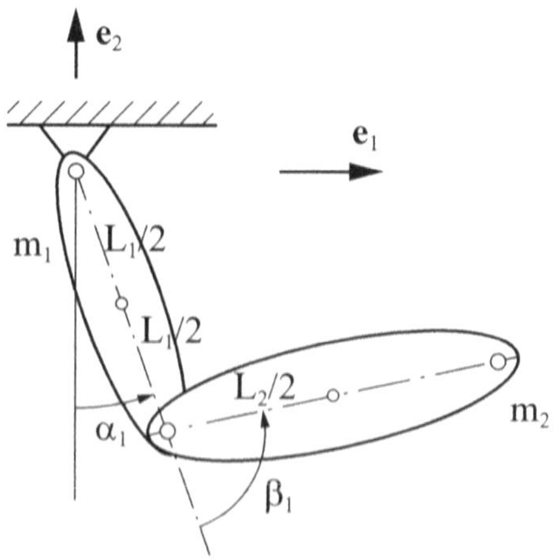

Fig. 5.16 Double pendulum

5.7.1 Non-recursive Formalisms

At this point we should briefly describe the symbolic formalism upon which Neweul-M^2 is based, see [30]. Neweul-M^2[1] is a research code based on Newton-Euler equations and the principles of d'Alembert and Jourdain. The first version was already developed in the 1970s at the University of Stuttgart by Kreuzer [29] and others. Neweul-M^2 generates, as far as is possible and sensible, equations of motion in minimal form (5.35) or (5.74), which can be combined with any code for the solution of ordinary differential equations. For very complex systems, recursive algorithms are also available in Neweul-M^2, and there are also many expansions, e.g. for the integration of flexible bodies. However, then the equations of motion can no longer be given in a fully symbolic way, and so only the basic algorithm will be clarified here using a simple example. Neweul-M^2 is based on Matlab with the Symbolic Math Toolbox (on the basis of Maple or Mupad) and can be utilized in various modi.

The double pendulum in Fig. 5.16 will be modeled and simulated in the following as a purposely simple example.

The use of a graphic user interface is especially helpful for the inexperienced user or newcomer, see Fig. 5.17. However, the GUI is also often used in order to construct a basic model that will later be more detailed in text files, and the user can switch as desired between the graphic user interface and the file interface. Figure 5.18 shows an easily comprehensible ASCII input file for Neweul-M^2. Any Matlab commands can be combined with the Neweul-M^2 commands, resulting in a highly powerful program environment.

Some explanations for the user: To begin with a new system is created by `newSys()`, and with `newGenCoord()` generalized coordinates and with

[1] For further information please consult www.itm.uni-stuttgart.de/research/neweul

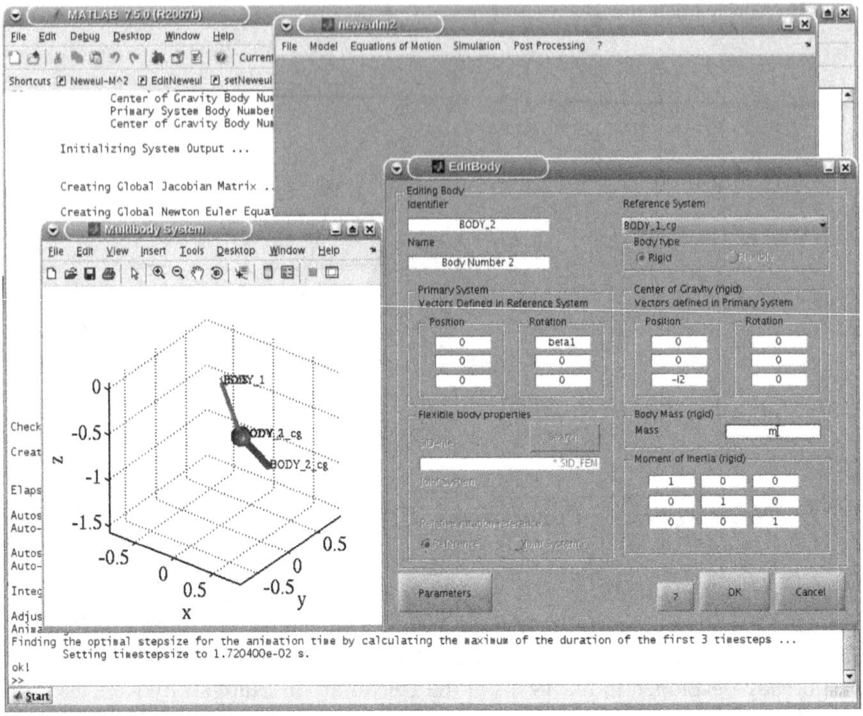

Fig. 5.17 Graphic user interface of Neweul-M^2

`newUserVarkonst()` variables are set. Then the bodies are defined with `newBody()`. It is possible here to mix symbolic and numerical quantities flexibly, and formulae can also be entered directly. Here the symbolic capabilities of Symbolic Math Toolbox are exploited. Then we can define force elements with `newForceElement()`, completing the description of the kinematics and kinetics. With the help of the methods described in the present volume, the nonlinear differential equations of motion can be set up with `calcEqMotNonLin()`, and the latter are then written with `writeMbsNonLin()`. These symbolic equations look like a manual calculation by the user, can be used for simulation in Matlab or exported for external use with other simulation programs. With `createAnimationWindow()` and `defineGraphics()`, a simple animation is prepared, yet we can also integrate complex CAD bodies into it as well. Now initial conditions can be introduced into the Matlab structure `sys.par.timeInt`, and `timeInt()` executes the time simulation. The results can now be viewed graphically, see Fig. 5.19, for a time plot of both state variables α_1 and β_1, or directly as a moving animation `animTimeInt()`, see Fig. 5.20.

```
% new system definition
newSys('Id','DP', 'Name','double pendulum',...
    'Gravity','[0; -g; 0]','frameOfReference','ISYS');

% generalized coordinates
newGenCoord('alpha1','beta1');

newUserVarKonst('l1',1,'b1',0.05,'m1',19.625, ...
    'l2',0.7,'b2',0.04,'m2',8.792,'T1',0,'T2',0);

% body defintions
newBody('Id','P1', ...
        'Name','Pendulum 1', ...
        'RefSys','ISYS', ...
        'RelRot','[0; 0; alpha1]', ...
        'CgPos','[l1/2; 0; 0]', ...
        'Mass','m1', ...
        'Inertia','[m1/12*(2*b1^2) 0 0; 0 m1/12*(l1^2+b1^2) 0; 0 0 m1/12*(l1^2+b1^2)]');
newBody('Id','P2', ...
        'Name','Pendulum 2', ...
        'RefSys','P1', ...
        'RelPos','[l1; 0; 0]', ...
        'RelRot','[0; 0; beta1]', ...
        'CgPos','[l2/2; 0; 0]', ...
        'Mass','m2', ...
        'Inertia','[m2/12*(2*b2^2) 0 0; 0 m2/12*(l2^2+b2^2) 0; 0 0 m2/12*(l2^2+b2^2)]');

% force elements
newForceElem('Id','FELEM_1', ...
             'Name','Force Element Pendulum 1', ...
             'Type','General', ...
             'Ksys1','P1', ...
             'Ksys2','ISYS', ...
             'DirDef','P1', ...
             'ForceLaw','[0; 0; 0; 0; 0; T1]');
newForceElem('Id','FELEM_2', ...
             'Name','Force Element Pendulum 2', ...
             'Type','General', ...
             'Ksys1','P2', ...
             'Ksys2','P1', ...
             'DirDef','P2', ...
             'ForceLaw','[0; 0; 0; 0; 0; T2]');

% create nonlinear equations of motion
calcEqMotNonLin;

% write functions for numerical evaluation
writeMbsNonLin;

% initialize animation
createAnimationWindow;
defineGraphics;     % model-specific!

% set initial conditions and run time integration
sys.par.timeInt.y0 = zeros(sys.dof,1);
sys.par.timeInt.Dy0 = zeros(sys.dof,1);
sys.par.timeInt.y0(1) = 56/180*pi;
sys.par.timeInt.y0(2) = 15/180*pi;
sys.results.timeInt = timeInt(sys.par.timeInt.y0, sys.par.timeInt.Dy0, ...
'Time',[0 10]);
animTimeInt;
```

Fig. 5.18 Neweul-M^2 input file

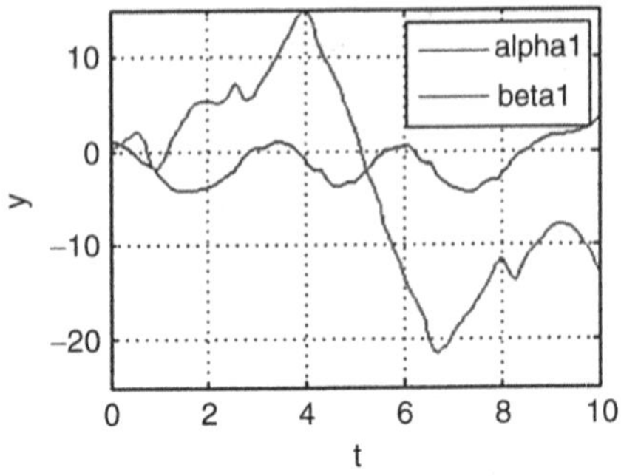

Fig. 5.19 Time plot of the generalized coordinates for the double pendulum

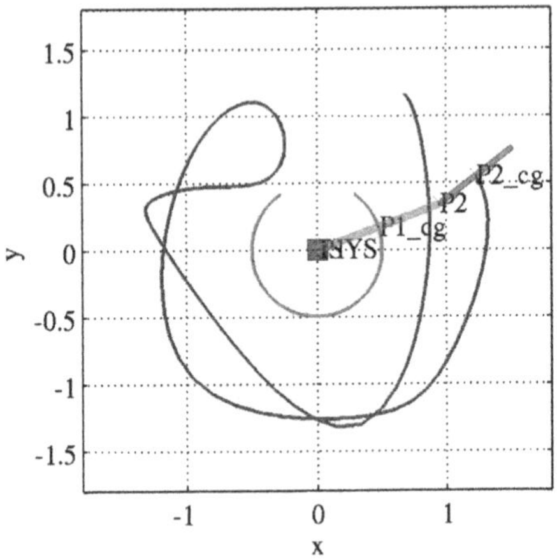

Fig. 5.20 Animation of the double pendulum

The output protocol of Neweul-M^2 can be seen in Fig. 5.21. We can clearly see the internal calculation sequence. After computing the kinematics, the global Newton-Euler equations are built and all necessary kinematic and kinetic variables are written. Then the required data for the time simulation are written.

As an example of typical Neweul-M^2 outputs, Fig. 5.22 shows the symbolic expressions for both vector coordinates of the generalized gyroscopic forces $\boldsymbol{k}$ and of the generalized forces $\boldsymbol{q}$ including the drive torques T_1, T_2 and the gravity forces.

For the optional calculation of all or some reaction forces, the Newton-Euler equations of the overall system are also available. From these we can obtain the purely algebraic equations of reaction (5.107) if needed.

5.7.2 *Recursive Formalisms*

When numerically solving the equations of motion (5.35) of large multibody systems in minimal form, the inversion of the often fully occupied inertia matrix proves to be complex. Hence, recursive methods have been developed, which avoid

```
Creating New System ... ok!
Generating Nonlinear Equations of Motion ...
Preparations ... ok!
Calculating Relative Kinematic Values  ok!
Calculating Absolute Kinematic Values ...
Inertial Frame ... ok!
Primary System Pendulum 1 ... ok!
Center of Gravity Pendulum 1 ... ok!
Primary System Pendulum 2 ... ok!
Center of Gravity Pendulum 2 ... ok!
ok!
Creating Global Jacobianan Matrix ... ok!
Creating Global Newton Euler Equations ...
Mass Matrix, Gen. Coriolis, Centrifugal and Gyroscopic Forces, Gravitation ...
Pendulum 1 ... ok!
Pendulum 2 ... ok!
Forces of Force Elements ...
Force Element Pendulum 1 ... ok!
Force Element Pendulum 2 ... ok!
Simplifying the Equations of Motion ...
Mass Matrix ... ok!
Local accelerations of frame of reference ... ok!
Coriolis and centrifugal forces ... ok!
Inner elastic forces ... ok!
ok!
Creating Functions for Numerical Evaluation ...
Auxiliary Functions ... ok!
Functions for the Coordinate Systems ... ok!
Functions for the System Dynamics ... ok!
Functions for the Equations of Motion ... ok!
ok!
Initializing the Animation ... ok!
Drawing coordinate systems ... ok!
Creating graphic objects ...
ok!
Animating Simulation results ...
Finding the optimal stepsize for the animation time by calculating
the maximum of the duration of the first 3 timesteps ...
Setting timestepsize to 1.974830e-02 s.
```

Fig. 5.21 Neweul-M^2 protocol file

```
sys.eqm.k
ans =
 -(Dbeta1*l1*l2*m2*sin(beta1)*(2*Dalpha1 + Dbeta1))/2
                   (Dalpha1^2*l1*l2*m2*sin(beta1))/2
sys.eqm.q
ans =
 T1 - (g*l2*m2*cos(alpha1 + beta1))/2 - (g*l1*m1*cos(alpha1))/2 - g*l1*m2*cos(alpha1)
                                                   T2 - (g*l2*m2*cos(alpha1 + beta1))/2
```

Fig. 5.22 Neweul-M^2 Matlab output

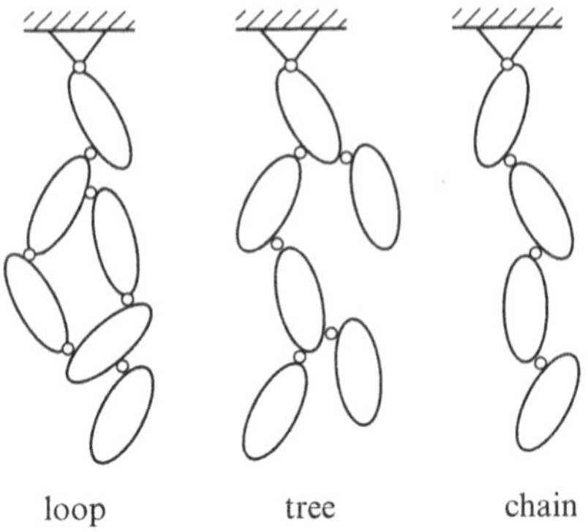

Fig. 5.23 Topology of multibody systems

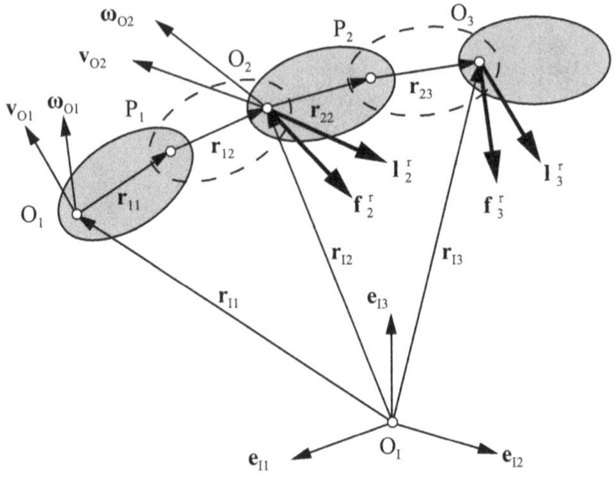

Fig. 5.24 Relative motion of two bodies

this inversion and thus enhance numerical efficiency, see Hollerbach [27], Bae and Haug [4], Brandl, Johanni, and Otter [11], Schiehlen [49]. The foundations of recursive formalisms will be introduced here for holonomic multibody systems.

The chain or tree topology of the multibody system under consideration is an important prerequisite for recursive formalisms, see Fig. 5.23.

A loop structure is not directly permissible. If loops arise nonetheless, then they are reduced into a tree topology by sectioning, and a few algebraic closing conditions must be observed. Recursive kinematics makes use of the relative motion between two neighboring bodies and associated local constraints, see Fig. 5.24.

According to (2.86), the absolute 6×1 twist $\boldsymbol{w}_i$ of the rigid body K_i with the body-fixed reference point O_i is composed of the 3×1 velocity vector $\boldsymbol{v}_{Oi}$ of the reference point and the 3×1 rotational velocity vector $\boldsymbol{\omega}_i$. The twist uniquely describes the state of velocity of the body K_i. The absolute twist $\boldsymbol{w}_i$ is now related to the absolute twist $\boldsymbol{w}_{i-1}$ of the previous body K_{i-1} and the generalized relative velocities $\dot{\boldsymbol{y}}_i$ in the joint P_{i-1} between both bodies. Then we can apply

$$\underbrace{\begin{bmatrix} \boldsymbol{v}_{Oi} \\ \boldsymbol{\omega}_i \end{bmatrix}}_{\boldsymbol{w}_i} = \underbrace{\boldsymbol{S}_{i,i-1} \cdot \begin{bmatrix} \boldsymbol{E} & -\tilde{\boldsymbol{r}}_{Oi-1,Oi} \\ \boldsymbol{0} & \boldsymbol{E} \end{bmatrix}}_{\boldsymbol{C}_i} \cdot \underbrace{\begin{bmatrix} \boldsymbol{v}_{Oi-1} \\ \boldsymbol{\omega}_{i-1} \end{bmatrix}}_{\boldsymbol{w}_{i-1}} + \underbrace{\boldsymbol{S}_{i,i-1} \cdot \begin{bmatrix} \boldsymbol{J}_{Ti} \\ \boldsymbol{J}_{Ri} \end{bmatrix}}_{\boldsymbol{J}_i} \cdot \dot{\boldsymbol{y}}_i, \tag{5.166}$$

where $\boldsymbol{S}_{i,i-1}$ is a 6×6 block diagonal matrix with two blocks of the relative 3×3 rotation tensor between the body-fixed frames i and $i-1$. The 6×6 matrix $\boldsymbol{C}_i$ is determined according (2.255) by the location vector between points O_{i-1} and O_i, while the $6 \times f_i$ matrix $\boldsymbol{J}_i$ is composed of the relative Jacobian matrices of translation and rotation. The sum $f = \sum_{i=1}^{p} f_i$ still applies for the total number of degrees of freedom of the system, also see (2.197). According to the rules of relative motion from Sect. 2.4, we obtain for the absolute acceleration in compliance with (2.257) and (2.258)

$$\boldsymbol{b}_i = \boldsymbol{C}_i \cdot \boldsymbol{b}_{i-1} + \boldsymbol{J}_i \cdot \ddot{\boldsymbol{y}}_i + \boldsymbol{\beta}_i(\dot{\boldsymbol{y}}_i, \boldsymbol{w}_{i-1}), \tag{5.167}$$

where the 6×1 vectors $\boldsymbol{b}_i$ and $\boldsymbol{b}_{i-1}$ summarize the translation and rotation accelerations, while the remaining velocity-dependent terms are found in the 6×1 vector $\boldsymbol{\beta}_i$. For the overall multibody system, the absolute acceleration can be given with the $6p \times 1$ vectors $\boldsymbol{b}$ and $\boldsymbol{\beta}$ as well as the $6p \times 6p$ matrices $\boldsymbol{C}$ and $\boldsymbol{J}$,

$$\boldsymbol{b} = \boldsymbol{C} \cdot \boldsymbol{b} + \boldsymbol{J} \cdot \ddot{\boldsymbol{y}} + \boldsymbol{\beta}. \tag{5.168}$$

The geometry matrix $\boldsymbol{C}$ is a lower block secondary diagonal matrix, while the Jacobian matrix $\boldsymbol{J}$ has a diagonal form

$$\boldsymbol{C} = \begin{bmatrix} \boldsymbol{0} & \boldsymbol{0} & \boldsymbol{0} & \cdots & \boldsymbol{0} \\ \boldsymbol{C}_2 & \boldsymbol{0} & \boldsymbol{0} & \cdots & \boldsymbol{0} \\ \boldsymbol{0} & \boldsymbol{C}_3 & \boldsymbol{0} & \cdots & \boldsymbol{0} \\ \vdots & \vdots & \ddots & \ddots & \vdots \\ \boldsymbol{0} & \boldsymbol{0} & \boldsymbol{0} & \boldsymbol{C}_p & \boldsymbol{0} \end{bmatrix}, \quad \boldsymbol{J} = \begin{bmatrix} \boldsymbol{J}_1 & \boldsymbol{0} & \boldsymbol{0} & \cdots & \boldsymbol{0} \\ \boldsymbol{0} & \boldsymbol{J}_2 & \boldsymbol{0} & \cdots & \boldsymbol{0} \\ \boldsymbol{0} & \boldsymbol{0} & \boldsymbol{J}_3 & \cdots & \boldsymbol{0} \\ \vdots & \vdots & \ddots & \ddots & \vdots \\ \boldsymbol{0} & \boldsymbol{0} & \boldsymbol{0} & \boldsymbol{0} & \boldsymbol{J}_p \end{bmatrix}. \tag{5.169}$$

From (5.168) we obtain the non-recursive form of the absolute acceleration

$$\boldsymbol{b} = (\boldsymbol{E} - \boldsymbol{C})^{-1} \cdot \boldsymbol{J} \cdot \ddot{\boldsymbol{y}} + \overline{\boldsymbol{\beta}}, \tag{5.170}$$

where the global $6p \times f$ Jacobian matrix $\overline{\boldsymbol{J}}$ appears again, also see (5.21),

$$\overline{\boldsymbol{J}} = (\boldsymbol{E} - \boldsymbol{C})^{-1} \cdot \boldsymbol{J} = \begin{bmatrix} \boldsymbol{J}_1 & \boldsymbol{0} & \boldsymbol{0} & \cdots & \boldsymbol{0} \\ \boldsymbol{C}_2 \cdot \boldsymbol{J}_1 & \boldsymbol{J}_2 & \boldsymbol{0} & \cdots & \boldsymbol{0} \\ \boldsymbol{C}_3 \cdot \boldsymbol{C}_2 \cdot \boldsymbol{J}_1 & \boldsymbol{C}_3 \cdot \boldsymbol{J}_2 & \boldsymbol{J}_3 & \cdots & \boldsymbol{0} \\ \vdots & \vdots & \ddots & \ddots & \vdots \\ * & * & * & \cdots & \boldsymbol{J}_p \end{bmatrix}. \tag{5.171}$$

Due to the chain topology of the multibody system, the global Jacobian matrix is a lower triangle matrix.

The Newton equation and the Euler equation are written for the body K_i in the body-fixed frame with the node O_i, where the absolute wrench $\boldsymbol{b}_i$ and the wrench $\boldsymbol{q}_i$ of the external forces and torques are used,

$$\underbrace{\begin{bmatrix} m_i\boldsymbol{E} & m_i\tilde{\boldsymbol{r}}_{OiCi}^T \\ m_i\tilde{\boldsymbol{r}}_{OiCi} & \boldsymbol{I}_{Oi} \end{bmatrix}}_{\boldsymbol{M}_i = \text{constant}} \cdot \underbrace{\begin{bmatrix} \boldsymbol{a}_{Oi} \\ \boldsymbol{\alpha}_{Oi} \end{bmatrix}}_{\boldsymbol{b}_i} + \underbrace{\begin{bmatrix} m_i\tilde{\boldsymbol{\omega}}_i \cdot \tilde{\boldsymbol{\omega}}_i \cdot \boldsymbol{r}_{OiCi} \\ \tilde{\boldsymbol{\omega}}_i \cdot \boldsymbol{I}_{Oi} \cdot \boldsymbol{\omega}_i \end{bmatrix}}_{\boldsymbol{k}_i} = \underbrace{\begin{bmatrix} \boldsymbol{f}_i \\ \boldsymbol{l}_{Oi} \end{bmatrix}}_{\boldsymbol{q}_i}. \tag{5.172}$$

Here, the 6×6 inertia matrix $\boldsymbol{M}_i$ is constant. The wrench $\boldsymbol{q}_i$ is now subdivided into an applied wrench and a wrench of the reactions, whereby the latter is determined by the reactions in the joints O_i and O_{i+1},

$$\boldsymbol{q}_i = \boldsymbol{q}_i^e + \boldsymbol{q}_i^r, \qquad \boldsymbol{q}_i^r = \boldsymbol{Q}_i \cdot \boldsymbol{g}_i - \boldsymbol{C}_{i+1}^T \cdot \boldsymbol{Q}_{i+1} \cdot \boldsymbol{g}_{i+1}. \tag{5.173}$$

Also, $\boldsymbol{Q}_i$ and $\boldsymbol{Q}_{i+1}$ are the $6 \times q_i$ distribution matrices in the corresponding joints, while the transposed geometry matrix $\boldsymbol{C}_{i+1}^T$ undertakes the transformation of the wrench $\boldsymbol{q}_{i+1}^{(r)}$ according to O_i, and the reaction law is applied. For the total number of system constraints, $q = \sum_{i=1}^{p} q_i$ still applies, see (2.197). According to (5.170), (5.172), and (5.173), we thus have $18p$ global equations available for the overall multibody system

$$\boldsymbol{b} = \overline{\boldsymbol{J}} \cdot \ddot{\boldsymbol{y}} + \overline{\boldsymbol{\beta}}, \tag{5.174}$$

$$\overline{\overline{\boldsymbol{M}}} \cdot \boldsymbol{b} + \overline{\boldsymbol{k}} = \boldsymbol{q}^{(e)} + \boldsymbol{q}^{(r)}, \tag{5.175}$$

$$\boldsymbol{q}^{(r)} = (\boldsymbol{E} - \boldsymbol{C})^T \cdot \boldsymbol{Q} \cdot \boldsymbol{g} = \overline{\boldsymbol{Q}} \cdot \boldsymbol{g} \tag{5.176}$$

with the $18p$ unknowns in the vectors $\boldsymbol{b}$, $\boldsymbol{y}$, $\boldsymbol{q}^{(r)}$, $\boldsymbol{g}$. The global, sparsely occupied distribution matrix $\overline{\boldsymbol{Q}}$ also appears, exhibiting only blocks on the diagonals and the upper secondary diagonal

$$\overline{\boldsymbol{Q}} = \begin{bmatrix} \boldsymbol{Q}_1 & -\boldsymbol{C}_2 \cdot \boldsymbol{Q}_2 & 0 & \cdots & 0 \\ 0 & \boldsymbol{Q}_2 & -\boldsymbol{C}_3 \cdot \boldsymbol{Q}_3 & \cdots & 0 \\ 0 & 0 & \boldsymbol{Q}_3 & \cdots & 0 \\ \vdots & \vdots & \ddots & \ddots & \vdots \\ 0 & 0 & 0 & 0 & \boldsymbol{Q}_p \end{bmatrix}. \tag{5.177}$$

If we now insert (5.174) and (5.176) into (5.175) and apply the orthogonality relation $\overline{\boldsymbol{J}}^T \cdot \overline{\boldsymbol{Q}} = \boldsymbol{0}$ according to (4.16), we then obtain the equations of motion in the known minimal form (5.35). The $f \times f$ inertia matrix is fully occupied however,

$$M = \begin{bmatrix} J_1^T\cdot(M_1+C_2^T\cdot(M_2+ +C_3^T\cdot M_3\cdot C_3)\cdot C_2)\cdot J_1 & J_1^T\cdot C_2^T(M_2+C_3^T\cdot M_3\cdot C_3)\cdot J_2 & J_1^T\cdot C_2^T\cdot C_3^T\cdot M_3\cdot J_3 \\ J_2^T\cdot(M_2+ +C_3^T\cdot M_3\cdot C_3)\cdot C_2\cdot J_1 & J_2^T\cdot(M_2+C_3^T\cdot M_3\cdot C_3)\cdot J_2 & J_2^T\cdot C_3^T\cdot M_3\cdot J_3 \\ J_3^T\cdot M_3\cdot C_3\cdot C_2\cdot J_1 & J_3^T\cdot M_3\cdot C_3\cdot J_2 & J_3^T\cdot M_3\cdot J_3 \end{bmatrix}, \quad (5.178)$$

and the $f \times 1$ vector $\boldsymbol{k}$ of gyroscopic and Coriolis forces depends not only on the generalized coordinates but also on the absolute velocities with the global twist $\boldsymbol{w}$

$$\boldsymbol{k} = \boldsymbol{k}(\boldsymbol{y}, \dot{\boldsymbol{y}}, \boldsymbol{w}). \quad (5.179)$$

Due to the chain topology of the multibody system, the inertia matrix (5.178) exhibits a characteristic structure, which can be directly evaluated with a Gauss transformation, see e.g. Bronstein et al. [12] or Spivak [61], and a recursion formula.

The result is a total of three steps needed to determine the generalized accelerations required in the integration of the equations of motion.

Step 1: Forward recursion in order to determine the absolute motion, beginning with the base body $i = 1$. The motion of the body $i = 0$ not belonging to the system must be known. Often, the inertial frame serves as the body $i = 0$ so that its absolute acceleration vanishes.

Step 2: Backward recursion, beginning with the final body $i = p$ with a Gauss transformation. The equations of motion are the outcome of this step

$$\hat{\boldsymbol{M}} \cdot \ddot{\boldsymbol{y}} + \hat{\boldsymbol{k}} = \hat{\boldsymbol{q}} \quad (5.180)$$

where the $f \times f$ inertia matrix $\hat{\boldsymbol{M}}$ is a lower triangle matrix

$$\hat{\boldsymbol{M}} = \begin{bmatrix} J_1^T\cdot\tilde{M}_1\cdot J_1 & 0 & 0 \\ J_2^T\cdot\tilde{M}_2\cdot C_2\cdot J_1 & J_2^T\cdot\tilde{M}_2\cdot J_2 & 0 \\ J_3^T\cdot\tilde{M}_3\cdot C_3\cdot C_2\cdot J_1 & J_3^T\cdot\tilde{M}_3\cdot C_3\cdot J_2 & J_3^T\cdot\tilde{M}_3\cdot J_3 \end{bmatrix}. \quad (5.181)$$

The blocks in (5.181) are obtained from the $f_i \times f_i$ recursion formula

$$\tilde{M}_{i-1} = \bar{M}_{i-1} + C_i^T\cdot(\tilde{M}_i - \tilde{M}_i\cdot J_i\cdot(J_i^T\cdot\tilde{M}_i\cdot J_i)^{-1}\cdot J_i^T\cdot\tilde{M}_i)\cdot C_i. \quad (5.182)$$

Since $f_i \le 5$ is valid for a multibody system with a chain structure due to the local constraints, only small matrices are to be inverted.

Step 3: Forward recursion in order to determine the generalized accelerations $\ddot{\boldsymbol{y}}$, beginning with $i = 1$.

As required, the generalized reaction forces can also be obtained without additional computations. It is self-evident that a recursive algorithm entails additional numerical complexity. For this reason, only if there are more than 8–10 bodies in the chain does recursion increase efficiency compared to direct matrix inversion. Expansions have also been suggested for loop topologies, see Bae and Haug [4] or Saha and Schiehlen [47]. In light of the prospect of increasing efficiency, recursive formalisms have begun to be incorporated in commercial programs as well.

Chapter 6
Finite Element Systems

Descriptively speaking, a finite element system is obtained by subdividing a nonrigid continuum into geometrically simple subdomains, which are connected at discrete nodes. A material law, such as Hooke's law for linear-elastic material, then leads to internal forces and torques which are reflected in the stiffness matrix of a single finite element. The nodes of the elements are linked by means of holonomic constraints, and external forces and torques can also act upon the nodes. Many details concerning the finite element method can be found e.g. in Wriggers [67], Bathe [6], or Zienkiewicz and Taylor [68].

The global equations of motion of the overall system are obtained by assembling all the local equations of motion of the finite elements. We will therefore first consider the local equations of motion. In the context of this book, it is neither possible nor desirable to introduce all of the great diversity of finite elements available today in detail. Rather, the basic ideas will be introduced for the tetrahedral element and explained in detail using the example of the beam element. The three-dimensional beam element includes the tension-compression rod, the torsion bar, and the planar beam as special cases. D'Alembert's principle is applied both in the determination of local equations of motion and in setting up the global system equations. Small deformations are assumed, such as common in linearized structural dynamics.

A beam system is a model composed of rigid bodies and finite beam elements. As a result of the large motions of the rigid bodies, the finite beam elements, too, carry out large motions with usually small deformations. Thus we must take into consideration the laws of relative motion as well. Finally, some information concerning the computation of strength for finite element systems will be provided.

6.1 Local Equations of Motion

The local equations of motion of a finite element system are valid for individual free elements that not are subject to external constraints. Without limiting generality, it is thus sufficient to consider one arbitrary finite element K. To establish the local

W. Schiehlen and P. Eberhard, *Applied Dynamics*, DOI 10.1007/978-3-319-07335-4_6,

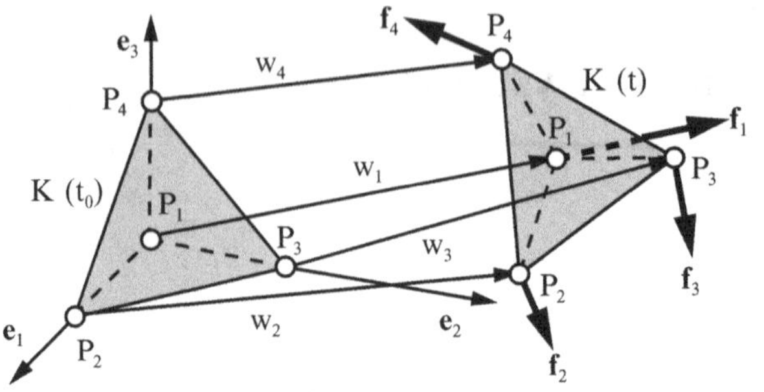

Fig. 6.1 Finite tetrahedral element

equations of motion, the momentum balance (3.63) is used along with the stress tensor given by Hooke's law and the concentrated forces acting at the nodes. Yet the momentum balance (3.63) cannot be evaluated since the deformation within the elastic element K is unknown.

To solve this problem, the finite element is now subjected to internal constraints by introducing shape functions. This prerequisite, seemingly quite arbitrary at first glance, has nevertheless proven effective in mechanics for decades. For example, the well-known Bernoulli hypothesis of beam deflection also involves an internal constraint: The beam cross-section should be planar and perpendicular to the beam axis. Simple, but already highly effective internal constraints can be obtained by assuming constant strain, which corresponds to a linear shape function. The local equations of motion will now be introduced for a tetrahedral element and a three-dimensional beam element.

6.1.1 *Tetrahedral Element*

According to Fig. 6.1, for the current configuration $K(t)$ of a tetrahedral element we have

$$\boldsymbol{r}(\boldsymbol{\rho},t) = \boldsymbol{\rho} + \boldsymbol{C}(\boldsymbol{\rho}) \cdot \boldsymbol{x}(t) \tag{6.1}$$

with the 12×1 position vector

$$\boldsymbol{x}(t) = \begin{bmatrix} \boldsymbol{w}_1 & \boldsymbol{w}_2 & \boldsymbol{w}_3 & \boldsymbol{w}_4 \end{bmatrix} \tag{6.2}$$

and the 3×12 Jacobian matrix

$$\boldsymbol{J}(\boldsymbol{\rho}) = \boldsymbol{C}(\boldsymbol{\rho}), \tag{6.3}$$

which corresponds to the matrix of the shape functions. The tetrahedral element thus has $e = 12$ degrees of freedom. From (6.1) we obtain for acceleration

$$\boldsymbol{a}(\boldsymbol{\rho},t) = \boldsymbol{C}(\boldsymbol{\rho}) \cdot \ddot{\boldsymbol{x}}(t) \tag{6.4}$$

and the virtual quantities are

$$\delta \boldsymbol{r} = \boldsymbol{C} \cdot \delta \boldsymbol{x}, \tag{6.5}$$

$$\delta \boldsymbol{e} = \boldsymbol{B} \cdot \delta \boldsymbol{x} = \mathscr{V} \cdot \boldsymbol{A} \cdot \delta \boldsymbol{x}. \tag{6.6}$$

If we also take Hooke's material law (3.68) into account, with the d'Alembert principle (4.31) the local equations of motion read as

$$\boldsymbol{M} \cdot \ddot{\boldsymbol{x}}(t) + \boldsymbol{K} \cdot \boldsymbol{x}(t) = \boldsymbol{q}(t). \tag{6.7}$$

Here,

$$\boldsymbol{M} = \rho \int_V \boldsymbol{C}^T \cdot \boldsymbol{C} dV \tag{6.8}$$

is the time-invariant 12×12 inertia matrix. The 12×12 stiffness matrix is also time-invariant,

$$\boldsymbol{K} = \int_V \boldsymbol{B}^T \cdot \boldsymbol{H} \cdot \boldsymbol{B} dV \tag{6.9}$$

and in the 12×1 vector $\boldsymbol{q}(t)$ the applied volume and surface forces are summarized as

$$\boldsymbol{q}(t) = \rho \int_V \boldsymbol{C}^T \cdot \boldsymbol{f} dV + \int_A \boldsymbol{C}^T \cdot \boldsymbol{t}^e dA. \tag{6.10}$$

The applied surface forces comprise the generalized forces acting on or being projected to the four nodes

$$\int_A \boldsymbol{C}^T \cdot \boldsymbol{t}^e dA = \begin{bmatrix} \boldsymbol{f}_1 & \boldsymbol{f}_2 & \boldsymbol{f}_3 & \boldsymbol{f}_4 \end{bmatrix}, \tag{6.11}$$

which are shown in Fig. 6.1. The elements of the matrices and vector will not be provided here, since they can be looked up in the literature, see e.g. Bathe [6]. The widely used linear shape functions $\boldsymbol{C} = \boldsymbol{D} + \boldsymbol{E} \cdot \boldsymbol{\rho}$ correspond to the internal constraints of a constant strain in the entire tetrahedral element.

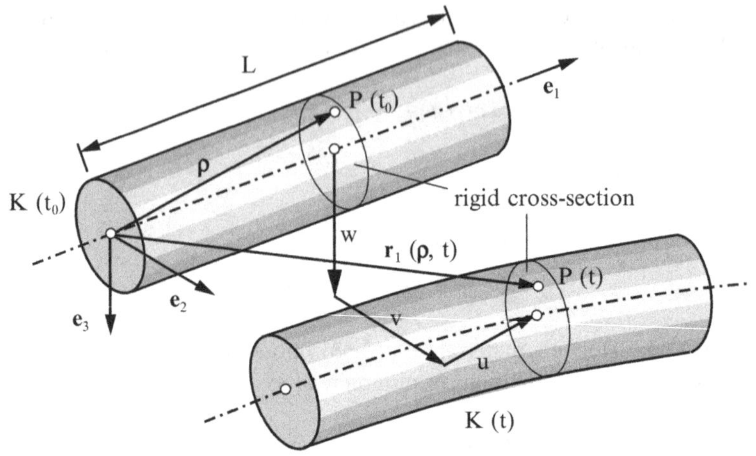

Fig. 6.2 Three-dimensional beam element

6.1.2 Three-Dimensional Beam Element

Because of its major engineering importance, the three-dimensional beam element, see Fig. 6.2, will now be treated in more detail. In engineering beam theory, one first assumes that the cross-sections of a beam remain plane and perpendicular to the beam axis. This leads to the internal constraints of a continuous beam, which have then the form

$$\boldsymbol{r}(\boldsymbol{\rho},t) = \begin{bmatrix} \rho_1 + u(\rho_1,t) + \beta(\rho_1,t)\rho_3 - \gamma(\rho_1,t)\rho_2 \\ \rho_2 + v(\rho_1,t) - \alpha(\rho_1,t)\rho_3 \\ \rho_3 + w(\rho_1,t) + \alpha(\rho_1,t)\rho_2 \end{bmatrix}. \tag{6.12}$$

Coordinates u, v and w denote the displacement of the beam axis $\boldsymbol{\rho} = [\rho_1 \quad 0 \ 0]$, while α, β, and γ represent the rotations of the beam cross-sections. Due to the orthogonality of the beam axis and the cross-section, it is also valid that

$$\beta = -\frac{\partial w}{\partial \rho_1} = -w', \qquad \gamma = \frac{\partial v}{\partial \rho_1} = v'. \tag{6.13}$$

Coordinate u is also called the longitudinal displacement as a result of tension-compression loading, while v and w describe deflection and α the torsional angle.

Every cross-section of the beam is thus described by one point on the beam axis with the coordinate ρ_1 and has six degrees of freedom. The beam axis is on the other hand a one-dimensional continuum with $a \to \infty$ degrees of freedom. As a result, the continuous beam exhibits according to (6.12) a total of $f = 6a$ with $a \to \infty$ degrees of freedom. For this reason, the beam can also be considered as a one-dimensional polar continuum. Because the theory of polar continua, or

Cosserat continua, respectively, has not been introduced, we will proceed only with the treatment of the beam as a nonpolar continuum with internal constraints.

The finite beam element is obtained from (6.12) by the additional introduction of internal constraints. For the longitudinal displacement and torsion, a linear function is possible such as it also arises in the solution of elastostatic problems. From

$$u(\rho_1,t) = c_1(t) + \rho_1 c_2(t) \tag{6.14}$$

after adjusting the boundary conditions to the beam ends,

$$u(0,t) = u_l(t) = c_1(t), \tag{6.15}$$

$$u(L,t) = u_r(t) = c_1(t) + Lc_2(t), \tag{6.16}$$

and normalizing the coordinate ρ_1, we get the shape function for the longitudinal displacement

$$u(x,t) = \begin{bmatrix} (1-x) & x \end{bmatrix} \cdot \begin{bmatrix} u_l(t) & u_r(t) \end{bmatrix}, \qquad x = \frac{\rho_1}{L}. \tag{6.17}$$

We also obtain for the torsion

$$\alpha(x,t) = \begin{bmatrix} (1-x) & x \end{bmatrix} \cdot \begin{bmatrix} \alpha_l(t) & \alpha_r(t) \end{bmatrix}, \qquad x = \frac{\rho_1}{L}. \tag{6.18}$$

On the other hand, at least a cubic function is required for the deflections due to the boundary conditions, since according to (6.13) the rotations β, γ are related to the deflections w, v, and four boundary conditions must be satisfied. With the function

$$v(\rho_1,t) = c_1(t) + \rho_1 c_2(t) + \rho_1^2 c_3(t) + \rho_1^3 c_4(t) \tag{6.19}$$

we thus find, after adjusting to the boundary conditions and normalizing the coordinate ρ_1, the shape function

$$v(x,t) = \begin{bmatrix} (1-x)^2(1+2x) \\ x(1-x)^2 L \\ x^2(3-2x) \\ -x^2(1-x)L \end{bmatrix} \cdot \begin{bmatrix} v_l(t) \\ \gamma_l(t) \\ v_r(t) \\ \gamma_r(t) \end{bmatrix}, \qquad x = \frac{\rho_1}{L}. \tag{6.20}$$

Analogous expressions are also found for $w(x,t)$, where the sign of the elements multiplied by L must be reversed because $w' = -\beta$.

The finite beam element thus has $e = 12$ degrees of freedom, which correspond to the rigid body motions of the cross-sections on the left and right beam ends. Since (6.17), (6.18), and (6.20) are decoupled however, the element matrices for the individual load cases can be defined independently of each other.

Equations (6.12) and (6.17) first yield the Jacobian matrix for the longitudinal displacement of the finite beam element

$$\boldsymbol{C}(x) = \begin{bmatrix} 1-x & x \\ 0 & 0 \\ 0 & 0 \end{bmatrix}. \tag{6.21}$$

The inertia matrix for the longitudinal displacement of the beam according to (6.8) thus is

$$\boldsymbol{M} = \rho A L \int_0^1 \begin{bmatrix} (1-x)^2 & (1-x)x \\ (1-x)x & x^2 \end{bmatrix} dx. \tag{6.22}$$

After evaluating the integrals, we obtain for the inertia matrix

$$\boldsymbol{M} = \frac{\rho A L}{6} \begin{bmatrix} 2 & 1 \\ 1 & 2 \end{bmatrix}. \tag{6.23}$$

Here, A is the constant cross-sectional area of the beam with mass $m = \rho A L$.

In order to calculate the stiffness matrix, first the strain and stress are determined. From (6.17), we get the one-dimensional state of strain

$$\boldsymbol{e}(t) = \begin{bmatrix} -\frac{1}{L} & \frac{1}{L} \\ 0 & 0 \\ 0 & 0 \\ 0 & 0 \\ 0 & 0 \\ 0 & 0 \end{bmatrix} \cdot \begin{bmatrix} u_l(t) \\ u_r(t) \end{bmatrix}, \tag{6.24}$$

which with Hooke's law (3.68) leads to a three-dimensional state of stress,

$$\boldsymbol{\sigma} = \frac{E}{(1-\nu)(1-2\nu)} \begin{bmatrix} -\frac{1-\nu}{L} & \frac{1-\nu}{L} \\ -\frac{\nu}{L} & \frac{\nu}{L} \\ -\frac{\nu}{L} & \frac{\nu}{L} \\ 0 & 0 \\ 0 & 0 \\ 0 & 0 \end{bmatrix}, \tag{6.25}$$

which contradicts experience. We can see that only for vanishing lateral strains, $\nu = 0$, do we have a one-dimensional state of stress. The reason for this contradiction is based on the fact that in (6.12) not only planar but also rigid cross-sections were assumed. In the case of the free beam element, we can overcome the contradiction by inserting additional terms into (6.12) that take into account the influence of lateral stress. Then there is a one-dimensional state of strain and a one-dimensional state of stress. However, then the beam element can no longer be connected over the entire cross-sectional area with a rigid body, since the geometric boundary conditions would be violated. A complete solution of this dilemma can only be found by means of a more exact modelling of the connection between a rigid body and the beam (e.g. via three-dimensional tetrahedral elements). In practice, the engineer can manage this problem either with a radially adjustable bearing between the beam and the rigid body or with an anisotropic material law. Both variants lead to a result that is compatible with experience (Saint Venant's principle). The anisotropic principle is used as a rule in the literature. It simplifies the calculation of element matrices thus will be made use of here as well.

Hence, instead of the matrix (3.69) of Hooke's law, in the case of the beam we have the relation

$$\boldsymbol{H} = \left[\begin{array}{ccc|ccc} 1+\nu & 0 & 0 & & & \\ 0 & 1+\nu & 0 & & \mathbf{0} & \\ 0 & 0 & 1+\nu & & & \\ \hline & & & \frac{1}{2} & 0 & 0 \\ & \mathbf{0} & & 0 & \frac{1}{2} & 0 \\ & & & 0 & 0 & \frac{1}{2} \end{array}\right] \frac{E}{1+\nu}. \tag{6.26}$$

The resulting stiffness matrix for the longitudinal displacement of the beam is written according to (6.9) as

$$\boldsymbol{K} = \frac{AE}{L}\int_0^1 \begin{bmatrix} 1 & -1 \\ -1 & 1 \end{bmatrix} dx. \tag{6.27}$$

Evaluated, we obtain for the stiffness matrix

$$\boldsymbol{K} = \frac{AE}{L}\begin{bmatrix} 1 & -1 \\ -1 & 1 \end{bmatrix}. \tag{6.28}$$

If we also assume a constant volume force density $\boldsymbol{f} = [n \quad 0 \quad 0]$ with the constant specific longitudinal load n, then the following applies for the generalized forces,

$$\begin{aligned} \boldsymbol{q}(t) &= \rho AL \int_0^1 \begin{bmatrix} 1-x & 0 & 0 \\ x & 0 & 0 \end{bmatrix} \cdot \begin{bmatrix} n \\ 0 \\ 0 \end{bmatrix} dx + \int_{A_l} \begin{bmatrix} 1 \\ 0 \end{bmatrix} t_1^e dA + \int_{A_r} \begin{bmatrix} 0 \\ 1 \end{bmatrix} t_1^e dA \\ &= \frac{\rho ALn}{2}\begin{bmatrix} 1 \\ 1 \end{bmatrix} + \begin{bmatrix} N_l(t) \\ N_r(t) \end{bmatrix}. \end{aligned} \tag{6.29}$$

Here, $N_{l,r}$ are the normal forces, which are based on the normal stresses on the left and right ends of the beam with cross-sections A_l and A_r.

Equations (6.12) and (6.18) yield the Jacobian matrix for the torsion of the finite beam element

$$\boldsymbol{C}(x,\rho_2,\rho_3) = \begin{bmatrix} 0 & 0 \\ -(1-x)\rho_3 & -x\rho_3 \\ (1-x)\rho_2 & x\rho_2 \end{bmatrix}. \tag{6.30}$$

For the torsion, the inertia matrix (6.8) assumes the form

$$\boldsymbol{M} = \rho L \int_0^1 \int_A \begin{bmatrix} (1-x)^2 & (1-x)x \\ (1-x)x & x^2 \end{bmatrix} (\rho_2^2+\rho_3^2)dA = \frac{\rho L J_P}{6} \begin{bmatrix} 2 & 1 \\ 1 & 2 \end{bmatrix}, \tag{6.31}$$

where the polar area moment of inertia J_P appears. Also, for the stiffness matrix of torsion according to (6.9) with (6.6), (6.26), and (6.30), we obtain

$$\boldsymbol{K} = \frac{E}{2(1+\nu)L} \int_0^1 \int_A \begin{bmatrix} 1 & -1 \\ -1 & 1 \end{bmatrix} (\rho_2^2+\rho_3^2)dA = \frac{GJ_P}{L} \begin{bmatrix} 1 & -1 \\ -1 & 1 \end{bmatrix}, \tag{6.32}$$

where the shear modulus $G = E/2(1+\nu)$ is used.

The moments $M_{Tl,r}$ of the shear stresses in the cross-sections on the left and right ends remain as generalized forces

$$\boldsymbol{q}(t) = \int_{A_l} \begin{bmatrix} 0 & \rho_3 & -\rho_2 \\ 0 & 0 & 0 \end{bmatrix} \cdot \begin{bmatrix} t_{1l}^e \\ t_{2l}^e \\ t_{3l}^e \end{bmatrix} dA + \int_{A_r} \begin{bmatrix} 0 & 0 & 0 \\ 0 & \rho_3 & -\rho_2 \end{bmatrix} \cdot \begin{bmatrix} t_{1r}^e \\ t_{2r}^e \\ t_{3r}^e \end{bmatrix} dA = \begin{bmatrix} M_{Tl}(t) \\ M_{Tr}(t) \end{bmatrix}. \tag{6.33}$$

From (6.12), (6.13), and (6.20) we find for the Jacobian matrix of the deflection v and for the rotation γ of the finite beam element

$$\boldsymbol{C}(x,\rho_2) = \begin{bmatrix} 6x(1-x)\rho_2 & -(1-4x+3x^2) & -6x(1-x)\frac{\rho_2}{L} & x(2-3x)\rho_2 \\ (1-x)^2(1+2x) & x(1-x)^2L & x^2(3-2x) & -x^2(1-x)L \\ 0 & 0 & 0 & 0 \end{bmatrix}. \tag{6.34}$$

Thus the inertia and stiffness matrix can be calculated as shown above. After a few intermediate calculations, we finally obtain for the inertia matrix of the deflection

$$\boldsymbol{M} = \frac{\rho AL}{420} \begin{bmatrix} 156 & 22L & 54 & -13L \\ 22L & 4L^2 & 13L & -3L^2 \\ 54 & 13L & 156 & -22L \\ -13L & -3L^2 & -22L & 4L^2 \end{bmatrix} + \frac{\rho J_3}{30L} \begin{bmatrix} 36 & 3L & -36 & 3L \\ 3L & 4L^2 & -3L & -L^2 \\ -36 & -3L & 36 & -3L \\ 3L & -L^2 & -3L & 4L^2 \end{bmatrix}, \tag{6.35}$$

where J_3 is the area moment of inertia with respect to the 3-axis of the cross-section. Also, the stiffness matrix of the deflection reads as

$$\boldsymbol{K} = \frac{EJ_3}{L^3}\begin{bmatrix} 12 & 6L & -12 & 6L \\ 6L & 4L^2 & -6L & 2L^2 \\ -12 & -6L & 12 & -6L \\ 6L & 2L^2 & -6L & 4L^2 \end{bmatrix}. \tag{6.36}$$

The generalized forces are determined using the lateral forces and bending torques on the left and right ends of the beam

$$\boldsymbol{q}(t) = \left[\, Q_{2l}(t) \;\; M_{3l}(t) \;\; Q_{2r}(t) \;\; M_{3r}(t) \,\right]. \tag{6.37}$$

Analogous matrices and vectors apply for the deflection w and rotation β.

Thus the local equations of motion are also available for a finite beam element with respect to small motions to the inertial frame. They are valid provided that the beam's longitudinal axis coincides with the 1-axis of the inertial frame. If this is not the case, then appropriate coordinate transformations must be carried out, see e.g. Link [33].

6.2 Global Equations of Motion

Finite elements serve for the static and dynamic analysis of engineering design, i.e. they have to be assembled to a global system. For this purpose, the free finite elements are subjected to external constraints. The external constraints are formulated in the generalized coordinates of the individual elements, i.e. in their node coordinates. This provides us with a system of f degrees of freedom. In structural dynamics, in which only small motions arise with respect to the inertial frame, it is advisable to express the constraints in the inertial frame as well.

If we now summarize the generalized coordinates of the global system with p finite elements in the $f \times 1$ position vector

$$\mathbf{y}(t) = \left[\, y_1 \;\; y_2 \;\; \dots \;\; y_f \,\right], \tag{6.38}$$

then the constraints can be expressed as follows,

$$\mathbf{x}_i(t) = \boldsymbol{I}_i \cdot \mathbf{y}(t), \qquad i = 1(1)p, \tag{6.39}$$

where $\mathbf{x}_i(t)$ is the $e_i \times 1$ position vector of the ith finite element, and $\boldsymbol{I}_i$ represents a usually constant $e_i \times f$ Jacobian matrix. The equations of motion of the overall system for linear finite elements then follow from the d'Alembert principle (4.24) in the form

$$\boldsymbol{M} \cdot \ddot{\mathbf{y}}(t) + \boldsymbol{K} \cdot \mathbf{y}(t) = \boldsymbol{q}(t) \tag{6.40}$$

with the $f \times f$ inertia matrix

$$\boldsymbol{M} = \sum_{i=1}^{p} \boldsymbol{I}_i^T \cdot \boldsymbol{M}_i \cdot \boldsymbol{I}_i, \tag{6.41}$$

the $f \times f$ stiffness matrix

$$\boldsymbol{K} = \sum_{i=1}^{p} \boldsymbol{I}_i^T \cdot \boldsymbol{K}_i \cdot \boldsymbol{I}_i \tag{6.42}$$

and the vector of the generalized forces

$$\boldsymbol{q} = \sum_{i=1}^{p} \boldsymbol{I}_i^T \cdot \boldsymbol{q}_i. \tag{6.43}$$

In structural dynamics, linear conservative vibration systems are generally obtained. However, often the effects of damping are also taken into consideration with the 'convenience hypothesis'. To this end, (6.40) is supplemented with a damping matrix $\boldsymbol{D}$, multiplied with the first derivative of the position vector $\mathbf{y}(t)$. In accordance with the Rayleigh hypothesis, we often assume for the damping matrix

$$\boldsymbol{D} = a\boldsymbol{M} + b\boldsymbol{K}, \qquad a, b = \text{constant}, \tag{6.44}$$

with which we can simultaneously diagonalize the matrices $\boldsymbol{M}$, $\boldsymbol{D}$ and $\boldsymbol{K}$, which corresponds to a modal damping of all eigenforms.

The 'lumped-mass method' is also well-established. In this case, one dispenses with a consistent calculation of the inertia matrix $\boldsymbol{M}$, and the mass m_i of the ith finite element is distributed to its nodes. As a result, the inertia matrix is always diagonal, which is an computational advantage if a large number of elements are involved.

Apart from that, it should be noted both that generation of matrices (6.41) and (6.42) and the solution of resultant equations are generally executed nowadays by commercial program systems. Nonetheless, every user of such programs should be familiar with their theoretical background in order to interpret the result correctly. For further information, see e.g. Bathe [6] or Wriggers [67].

The equations of motion (6.40) also contain the elastostatics of supporting structures. With $\ddot{\mathbf{y}}(t) = \dot{\mathbf{y}}(t) = \mathbf{0}$, we can calculate the static deformation of a structure,

$$\mathbf{y}_{\text{stat}} = \boldsymbol{K}^{-1} \cdot \boldsymbol{q}. \tag{6.45}$$

The matrix inversion required for this can be carried out efficiently by exploiting the band structure of the stiffness matrix.

Fig. 6.3 Elastic truss

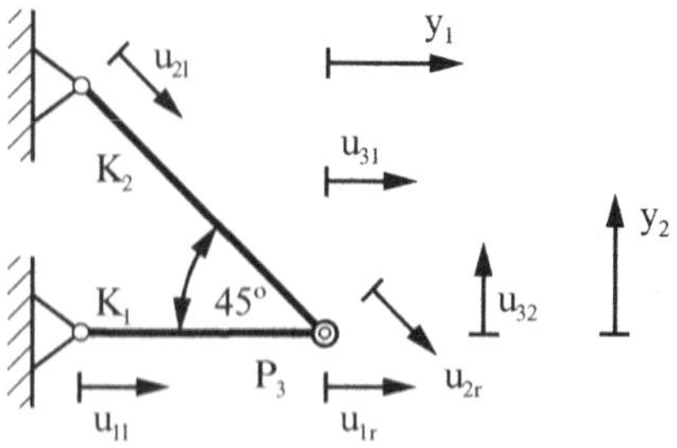

Example 6.1 (Truss Vibrations). The elastic truss shown in Fig. 6.3 consists of two bars K_1 and K_2 (length L and $\sqrt{2}L$, cross-section A, Young's modulus E) and a mass point P_3 (mass M), so a total of $p = 3$ elements. The mass of the bars in comparison to the mass M of the point can be neglected. If we dispense with writing the reaction forces, the local equations of motion of the sectioned system are

$$\frac{AE}{L}\begin{bmatrix} 1 & -1 \\ -1 & 1 \end{bmatrix} \cdot \begin{bmatrix} u_{1l} \\ u_{1r} \end{bmatrix} = \mathbf{0}, \tag{6.46}$$

$$\frac{AE}{\sqrt{2}L}\begin{bmatrix} 1 & -1 \\ -1 & 1 \end{bmatrix} \cdot \begin{bmatrix} u_{2l} \\ u_{2r} \end{bmatrix} = \mathbf{0}, \tag{6.47}$$

$$M\begin{bmatrix} 1 & 0 \\ 0 & 1 \end{bmatrix} \cdot \begin{bmatrix} \ddot{u}_{31} \\ \ddot{u}_{32} \end{bmatrix} = \begin{bmatrix} 0 \\ -Mg \end{bmatrix}. \tag{6.48}$$

The constraints are given by

$$\boldsymbol{x} = \begin{bmatrix} u_{1l} \\ u_{1r} \\ u_{2l} \\ u_{2r} \\ u_{31} \\ u_{32} \end{bmatrix} = \begin{bmatrix} 0 & 0 \\ 1 & 0 \\ 0 & 0 \\ 1/\sqrt{2} & -1/\sqrt{2} \\ 1 & 0 \\ 0 & 1 \end{bmatrix} \cdot \begin{bmatrix} y_1 \\ y_2 \end{bmatrix}. \tag{6.49}$$

In accordance with (6.41)–(6.43) we find the equations of motion

$$M\begin{bmatrix} 1 & 0 \\ 0 & 1 \end{bmatrix} \cdot \begin{bmatrix} \ddot{y}_1 \\ \ddot{y}_2 \end{bmatrix} + \frac{AE}{2\sqrt{2}L}\begin{bmatrix} 2\sqrt{2}+1 & -1 \\ -1 & 1 \end{bmatrix} \cdot \begin{bmatrix} y_1 \\ y_2 \end{bmatrix} = \begin{bmatrix} 0 \\ -Mg \end{bmatrix}. \tag{6.50}$$

We then obtain for the static excursion

$$\boldsymbol{y}_{stat} = \frac{L}{AE}\begin{bmatrix} 1 & 1 \\ 1 & 1+2\sqrt{2} \end{bmatrix} \cdot \begin{bmatrix} 0 \\ -Mg \end{bmatrix} = -\frac{MgL}{AE}\begin{bmatrix} 1 \\ 1+2\sqrt{2} \end{bmatrix}. \tag{6.51}$$

The mass point thus tends to the bottom left under the effect of its weight.

End of Example 6.1.

The finite element method is by definition an approximation method. But it can be shown that this method converges toward the exact solution provided that the number of finite elements is sufficiently large. Also, monotonic convergence is ensured when conformable shape functions are used that guarantee a continuous deformation across the element boundaries. The linear shape functions of the tetrahedral element and the shape function (6.19) for beam deflection agree with this conformity requirement. The accuracy of the result can therefore be tested, among other ways, by modelling with a varying number of finite elements.

6.3 Flexible Multibody Systems

In engineering dynamics, multibody systems are often encountered which are composed of rigid and elastic bodies. For example, industrial robots may have to be modelled as flexible multibody systems: the joints are treated as rigid bodies, and the links as elastic beams. As opposed to structural dynamics, beams in machine dynamics carry out large rigid body motions overlaid with elastic strains. It is thus advisable to introduce elastic coordinates in addition to the rigid body coordinates. In the standard case of small elastic strains a floating reference frame is chosen describing the large rigid body motions and the small elastic displacements of the nodes relative to the reference frame. In the case of large elastic strains the displacements of the nodes have to be described with respect to the inertial frame including the rigid body motion. Both formulations are explained in this section for a planar flexible beam element. Then, more details are presented for planar beam systems. But first of all we will show that the element matrices introduced in Sect. 6.1 contain the motions of rigid bodies after adding appropriate internal constraints.

The deformation of a rigid beam element, see Fig. 6.4, is obtained from (6.20) with the additional internal constraints

$$v_r = v_l + L\gamma_l, \qquad \gamma_r = \gamma_l, \qquad \gamma_l \ll 1 \tag{6.52}$$

or

$$v(x,t) = \begin{bmatrix} 1 & xL \end{bmatrix} \cdot \begin{bmatrix} v_l(t) \\ \gamma_l(t) \end{bmatrix}. \tag{6.53}$$

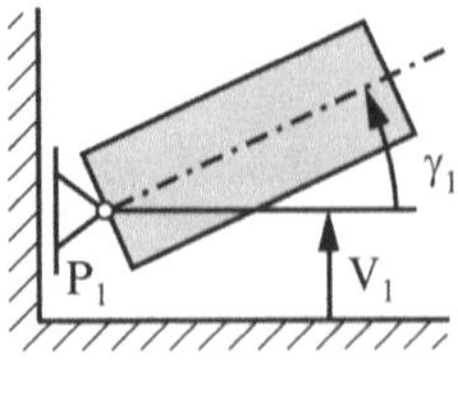

Fig. 6.4 Rigid beam element

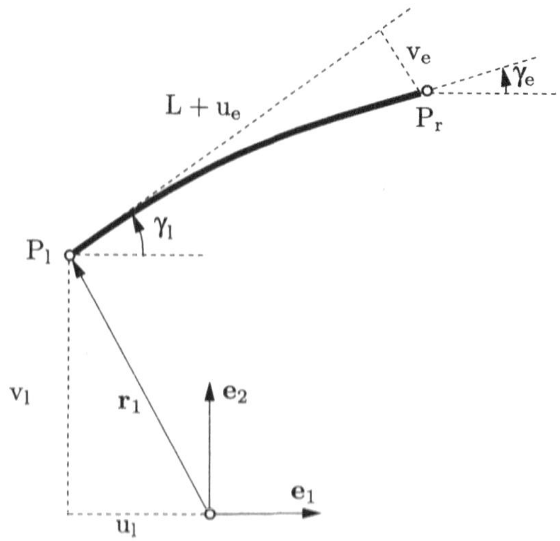

Fig. 6.5 Relative coordinates for a planar flexible beam element

Thus, (6.35) and (6.36) yield the element matrices

$$\boldsymbol{M} = \frac{m}{6}\begin{bmatrix} 6 & 3L \\ 3L & 2L^2 + 6N^2 \end{bmatrix}, \qquad \boldsymbol{K} = \boldsymbol{0}, \tag{6.54}$$

with the abbreviation $N^2 = \rho J_3 L/m$ where $m = \rho AL$ is the mass of the rigid beam with cross-section A and length L, and N denotes the inertia based on the local rotating mass of a cross section with area moments of inertia J_3. Moreover, the generalized applied forces read as

$$\boldsymbol{q}(t) = \begin{bmatrix} Q_l(t) + Q_r(t) & M_l(t) + M_r(t) + LQ_r(t) \end{bmatrix}. \tag{6.55}$$

These matrices and vectors determine the equations of the planar motion of a rigid beam, which can also be obtained by the principle of linear and angular momentum with respect to the node P_l.

6.3.1 Floating Frame of Reference Formulation

In particular, for the analysis of small structural vibrations the Floating Frame of Reference Formulation (FFRF) has been well established in engineering, see Schwertassek and Wallrapp [51], Geradin and Cardona [22], Bauchau [7], Seifried [52], and Shabana [57]. This modelling approach is explained by Fig. 6.5 for a planar beam element.

The large displacement rigid body motions are characterized by a body-fixed frame at the left node P_l of the beam and elastic deformations described by relative coordinates. Thus, the large rigid body motion is represented by the position vector of the left node P_l as

$$\boldsymbol{x}_s = [u_l \; v_l \; \gamma_l] \tag{6.56}$$

while the small elastic deformations read as

$$\boldsymbol{x}_e = [u_e \; v_e \; \gamma_e], \quad u_e \ll L, \quad v_e \ll L, \quad \gamma_e \ll 1. \tag{6.57}$$

These two position vectors describe the $f = 6$ nodal coordinates according to the 6 degrees of freedom of the elastic beam element. Then, the resulting global equations of motion are obtained as

$$\begin{bmatrix} \boldsymbol{M}_s(\boldsymbol{x}_s) & \boldsymbol{M}_{se}(\boldsymbol{x}) \\ \boldsymbol{M}_{es}(\boldsymbol{x}) & \boldsymbol{M}_e \end{bmatrix} \cdot \begin{bmatrix} \ddot{\boldsymbol{x}}_s(t) \\ \ddot{\boldsymbol{x}}_e(t) \end{bmatrix} + \begin{bmatrix} \boldsymbol{0} & \boldsymbol{0} \\ \boldsymbol{0} & \boldsymbol{K}_e \end{bmatrix} \cdot \begin{bmatrix} \boldsymbol{x}_s(t) \\ \boldsymbol{x}_e(t) \end{bmatrix} + \boldsymbol{k}(\boldsymbol{x},\dot{\boldsymbol{x}}) = \boldsymbol{q}(t). \tag{6.58}$$

In particular, the elastic stiffness matrix $\boldsymbol{K}$ is constant and it is usually available from finite element modelling of the flexible body under consideration. However, due to the relative coordinates chosen, the inertia matrix is very complex depending on some or all nodal coordinates which serve in FFRF as generalized coordinates. For the inertia matrix $\boldsymbol{M}(\boldsymbol{x})$ special inertia shape integrals are required as shown, e.g., by Schwertassek and Wallrapp [51] and Shabana [57] in detail. These inertia shape integrals can be pre-computed prior to the simulation what means a great computational advantage. Furthermore, the vector $\boldsymbol{k}(\boldsymbol{x},\dot{\boldsymbol{x}})$ represents the Coriolis and gyroscopic forces due to the large rigid body motion, and the vector $\boldsymbol{q}(t)$ summarizes the generalized applied forces.

6.3.2 Absolute Nodal Coordinate Formulation

The Floating Frame of Reference Formulation is most useful for engineering applications where usually the rigid body motion represents the design task of the system, and the structural vibrations are disturbances. By design and the chosen modes of operation the structural vibration of the system remain usually small as required. Nevertheless, there are systems with lightweight components and reduced stiffness like the rotating blades of helicopters, see Bauchau, Bottasso and Nikishkov [8]. In such cases the FFRF does not allow exact modelling even for the rigid body motion as pointed out by Shabana [54] and [55]. To overcome this problem Shabana proposed the Absolute Nodal Coordinate Formulation (ACNF) which is now explained with respect to a flexible beam element shown in Fig. 6.6 subject to planar motion.

The overall motion of a flexible beam is characterized by two nodes, the left one P_l and the right one P_r, in the inertial frame avoiding the incremental finite element

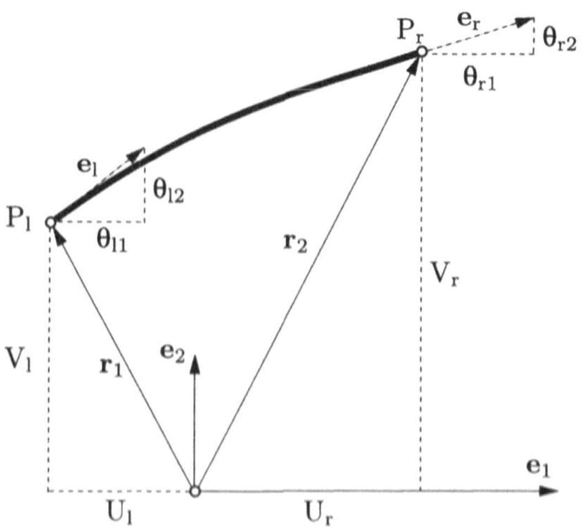

Fig. 6.6 Absolute coordinates for a planar flexible beam element

approach. Thus, the total motion including the exact rigid body motion and large elastic deformations is represented by an 8×1 position vector

$$\boldsymbol{e}(t) = [U_l \; V_l \; \Theta_{l1} \; \Theta_{l2} \; U_r \; V_r \; \Theta_{r1} \; \Theta_{r2}] \, . \tag{6.59}$$

In particular, the rotation of the slope at the nodes characterized by the unit vectors $\boldsymbol{e}_l$ and $\boldsymbol{e}_r$ is not restricted to small angles. Therefore, the rigid body motion is always modelled correctly. The coordinates of the position vector (6.59) are not independent from each other. With two constraints in mind, e.g.,

$$\Theta_{l1}^2 + \Theta_{l2}^2 = 1, \quad \Theta_{r1}^2 + \Theta_{r2}^2 = 1, \tag{6.60}$$

the position vector (6.59) describes the $f = 6$ degrees of freedom of a planar flexible beam element. The extended equations of motion read as

$$\boldsymbol{M} \cdot \ddot{\boldsymbol{e}}(t) + \boldsymbol{K}(\boldsymbol{e}) \cdot \boldsymbol{e}(t) = \boldsymbol{Q}(t). \tag{6.61}$$

In particular, the inertia matrix $\boldsymbol{M}$ is now constant while the stiffness matrix $\boldsymbol{K}(\boldsymbol{e})$ is highly nonlinear. For the evaluation of the stiffness matrix cubic global shape functions have to be used, and the stiffness matrices from the classical finite element approach cannot be used any longer. The computation of the nonlinear stiffness matrix is outlined by Shabana [57]. But for the simulation of motion a constant inertia matrix is most valuable since the inversion of the inertia matrix has to be computed only once. The price to be paid for is the additional effort with the stiffness matrix.

Both formulations, FFRF and ACND, have been compared by Escalona et al. [19] with respect to a crank slider mechanism which exhibits by design only small structural vibrations. The position of the slider and the deformation of the midpoint of the connecting rod show excellent agreement. But numerical experiments verify discrepancies between solutions by FFRF and ANCF in the case of large deformations.

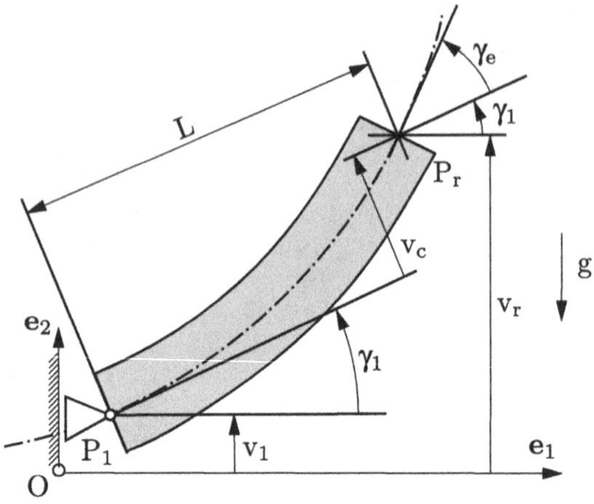

Fig. 6.7 Rigid body motion of an elastic beam element

6.3.3 Planar Beam Systems

For the discussion of beam systems we restrict ourselves to inductile beam elements with vanishing longitudinal displacement in the vertical plane e_1, e_2 subject to gravity as shown in Fig. 6.7. Then, the elastic coordinates v_e and γ_e remain, see Fig. 6.7, and the following applies for the coordinates on the right beam end P_r

$$v_r = v_l + L \sin\gamma_l + v_e, \qquad \gamma_r = \gamma_l + \gamma_e. \tag{6.62}$$

Furthermore, the internal constraints (6.12), i.e., $u = w = \alpha = \beta = 0$, assume the following form in the case of planar deflection in compliance with (2.134) and (6.34)

$$\boldsymbol{r}(\boldsymbol{\rho},t) = \underbrace{\begin{bmatrix} 0 \\ 1 \\ 0 \end{bmatrix}}_{\boldsymbol{e}_2} v_l + \boldsymbol{S}(\gamma_l) \cdot \left\{ \underbrace{\begin{bmatrix} \rho_1 \\ \rho_2 \\ \rho_3 \end{bmatrix}}_{\boldsymbol{\rho}} \right.$$

$$\left. + \underbrace{\begin{bmatrix} -6x(1-x)\rho_2/L & x(2-3x)\rho_2 \\ x^2(3-2x) & -x^2(1-x)L \\ 0 & 0 \end{bmatrix}}_{\boldsymbol{A}_e(x,\rho_2)} \cdot \underbrace{\begin{bmatrix} v_e \\ \gamma_e \end{bmatrix}}_{\boldsymbol{x}_e} \right\}, \qquad \rho_1 = xL. \tag{6.63}$$

We can see that (6.63) agrees with (6.12), (6.13), and (6.20) for $\gamma_l \ll 1$. The Jacobian matrix for large motions v_l, γ_l is now obtained under the assumption of small elastic deformations

$$|v_e| \ll L, \qquad |\gamma_e| \ll 1 \tag{6.64}$$

in the form

$$\boldsymbol{C}(x, \rho_2, \gamma_l) = \left[\; \boldsymbol{e}_2 \;\middle|\; \frac{\partial \boldsymbol{S}}{\partial \gamma_l} \cdot \boldsymbol{\rho} \;\middle|\; \boldsymbol{S} \cdot \boldsymbol{A}_e \; \right]. \tag{6.65}$$

After a rather elaborate intermediate calculation, we find the inertia matrix and stiffness matrix for the case of slim beams $|\rho_2|, |\rho_3| \ll \rho_1$ or $J_3 \ll m/\rho L^3$ with the position vector

$$\boldsymbol{x}(t) = \left[\; v_l \;\; \gamma_l \;\middle|\; v_e \;\; \gamma_e \; \right] \tag{6.66}$$

in the form

$$\boldsymbol{M} = \frac{m}{420} \left[\begin{array}{cc|cc} 420 & 210L\cos\gamma_l & 210\cos\gamma_l & -35L\cos\gamma_l \\ 210L\cos\gamma_l & 140L^2 & 147L & -21L^2 \\ \hline 210L\cos\gamma_l & 147L & 156 & -22L \\ -35L\cos\gamma_l & -21L^2 & -22L & 4L^2 \end{array} \right] \tag{6.67}$$

and

$$\boldsymbol{K} = \frac{EJ_3}{L^3} \left[\begin{array}{cc|cc} 0 & 0 & 0 & 0 \\ 0 & 0 & 0 & 0 \\ \hline 0 & 0 & 12 & -6L \\ 0 & 0 & -6L & 4L^2 \end{array} \right]. \tag{6.68}$$

In addition, a Coriolis force also appears

$$\boldsymbol{k} = \frac{m}{12} \left[\; (-6L\dot{\gamma}_l - 6\dot{v}_e + L\dot{\gamma}_e) \;\; 0 \;\middle|\; 0 \;\; 0 \; \right] \dot{\gamma}_l \sin\gamma_l \tag{6.69}$$

and the generalized weight is

$$\boldsymbol{q} = -\frac{mg}{12} \left[\; 12 \;\; 6L\cos\gamma_l \;\middle|\; 6\cos\gamma_l \;\; -L\cos\gamma_l \; \right]. \tag{6.70}$$

We first recognize the fact that the rigid body motions are coupled nonlinearly with the elastic motions via the Coriolis force according to (6.69). Also, there is a linear coupling of the inertia forces due to the fully occupied inertia matrix (6.67).

Only the stiffness matrix (6.68) of the elastic forces is decoupled by definition. In addition, the weights in (6.70) act upon all the motions. This reveals several new phenomena compared to structural mechanics. It should be noted in particular that the condition (6.64) must be satisfied by a sufficient amount of beam stiffness, $mg \ll EJ_3/L^2$ or $m\dot{\gamma}^2 \ll EJ_3/L^3$, respectively, since the existing approach, limited to the 2-direction, would otherwise entail destabilizing centrifugal forces. In the case of low beam bending stiffness, the motion coupling in the 1,2 plane must be taken into account. This coupling is based either on a kinematic constraint as a result of the inelastic beam axis assumed, or it is due to the elastic coupling between longitudinal displacement and bending. Accurate modelling of an elastic system is thus much more difficult than dealing with a rigid multibody system.

We thus obtain as a result the local equations of motion of a elastic beam element in nonlinear motion in a form which has been expanded with respect to (5.2) and (6.7)

$$\boldsymbol{M}(\boldsymbol{x}) \cdot \ddot{\boldsymbol{x}}(t) + \boldsymbol{K} \cdot \boldsymbol{x}(t) + \boldsymbol{k}(\boldsymbol{x}, \dot{\boldsymbol{x}}) = \boldsymbol{q}(t). \tag{6.71}$$

The vector $\boldsymbol{k}$ of the Coriolis or gyroscopic forces is based on large rigid body motions, while the stiffness matrix $\boldsymbol{K}$ denotes small elastic deformations.

The individual rigid bodies and the finite elements of a beam system must now be assembled into a global system. For this purpose, the local $e_i \times 1$ position vectors $\boldsymbol{x}_i$ are described by the $f \times 1$ position vector $\boldsymbol{y}(t)$ of the global system. The Jacobian matrices $\boldsymbol{I}_i$ appearing according to (6.39) are in general no longer constant.

If we further subdivide the $f \times 1$ position vector $\boldsymbol{y}(t)$ again into a position vector $\boldsymbol{y}_s(t)$ of rigid body motions and a position vector $\boldsymbol{y}_e(t)$ of elastic vibrations,

$$\boldsymbol{y}(t) = [\boldsymbol{y}_s \quad \boldsymbol{y}_e], \tag{6.72}$$

this yields, applying (6.41)–(6.43), the global equations of motion

$$\begin{bmatrix} \boldsymbol{M}_s(\boldsymbol{y}_s) & \boldsymbol{M}_{se}(\boldsymbol{y}) \\ \boldsymbol{M}_{es}(\boldsymbol{y}) & \boldsymbol{M}_e \end{bmatrix} \cdot \begin{bmatrix} \ddot{\boldsymbol{y}}_s(t) \\ \ddot{\boldsymbol{y}}_e(t) \end{bmatrix} + \begin{bmatrix} \boldsymbol{0} & \boldsymbol{0} \\ \boldsymbol{0} & \boldsymbol{K}_e \end{bmatrix} \cdot \begin{bmatrix} \boldsymbol{y}_s(t) \\ \boldsymbol{y}_e(t) \end{bmatrix} + \boldsymbol{k} = \boldsymbol{q}. \tag{6.73}$$

In order to solve the global equations of motion (6.73), we can get in two steps an approximation. First the large, generally nonlinear rigid body motions are computed. Then the linear elastic vibrations are determined with the rigid body motion as a given target motion. However, it must be observed that the linear differential equations may have time-variant coefficients and are subject to additional external excitations as a result of rigid body motions.

Example 6.2 (Beam Pendulum). The beam pendulum shown in Fig. 6.8 consists of two bodies, a simply supported slim beam (mass m_1, length L, bending stiffness EJ_3), and an attached mass point (mass m_2).

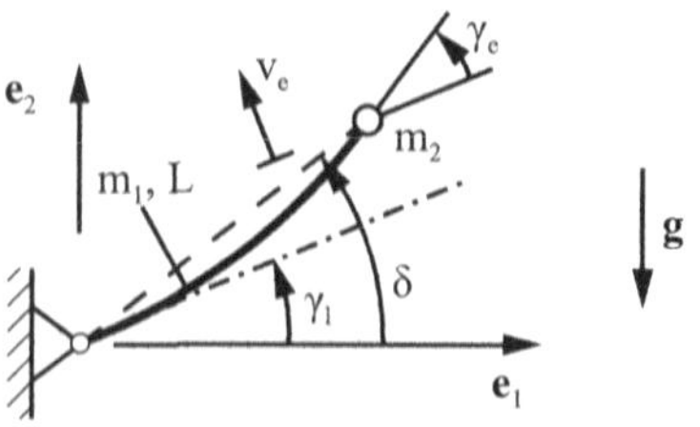

Fig. 6.8 Elastic beam pendulum

The local equations of motion of the beam are given by (6.71) with (6.66)–(6.70) subject to the constraint $v_l = 0$. Thus, the local position vector reads

$$\boldsymbol{x}_1(t) = [0 \quad \gamma_l \quad v_e \quad \gamma_e]\,. \tag{6.74}$$

The mass point m_2 moves on a circular path because the beam length can be considered as constant due to the small deflection. Thus the following equation applies for the motion of the mass point

$$m_2 L^2 \ddot{\delta}(t) = -m_2 g L \cos\delta(t) + LQ_r \tag{6.75}$$

with the angle $\delta(t)$ introduced in Fig. 6.8 and the reaction force Q_r at the right end of the beam element. Thus, for the scalar position variable of the mass point it remains the constraint

$$x_2(t) = \delta(t) = \gamma_l(t) + \frac{1}{L} v_e(t). \tag{6.76}$$

The beam system has $f = 3$ degrees of freedom described by the 3×1 position vector

$$\mathbf{y}(t) = \begin{bmatrix} \gamma_l & v_e & \gamma_e \end{bmatrix}, \quad |v_e| \ll 1, \quad |\gamma_e| \ll 1. \tag{6.77}$$

With the constraints given in Fig. 6.8 and in (6.74) and (6.76), the Jacobian matrices are then written for the bodies as according to (6.39)

$$\boldsymbol{I}_1 = \begin{bmatrix} 0 & 0 & 0 \\ 1 & 0 & 0 \\ 0 & 1 & 0 \\ 0 & 0 & 1 \end{bmatrix}, \qquad \boldsymbol{I}_2 = \begin{bmatrix} 1 & \frac{1}{L} & 0 \end{bmatrix}. \tag{6.78}$$

The global equations of motion thus assume the form (6.73), too. In particular, the equation of motion of the rigid body motion for $v_e = \gamma_e = 0$ is found to be

$$\frac{1}{3}(m_1 + 3m_2)L^2 \ddot{\gamma}_l + \frac{1}{2}(m_1 + 2m_2) g L \cos\gamma_l = 0. \tag{6.79}$$

The elastic vibrations lead to the equations of motion

$$\frac{m_1}{210}\begin{bmatrix} 78+210\frac{m_2}{m_1} & -11L \\ -11L & 2L^2 \end{bmatrix}\cdot\begin{bmatrix}\ddot{v}_e\\ \ddot{\gamma}_e\end{bmatrix}+\frac{EJ_3}{L^3}\begin{bmatrix} 12 & -6L \\ -6L & 4L^2 \end{bmatrix}\cdot\begin{bmatrix}v_e\\ \gamma_e\end{bmatrix}$$
$$=-\frac{m_1 g}{12}\cos\gamma_l\begin{bmatrix}6+12\frac{m_2}{m_1}\\ -L\end{bmatrix}-\frac{m_1 L}{420}\begin{bmatrix}147+420\frac{m_2}{m_1}\\ -21L\end{bmatrix}\ddot{\gamma}_l$$
$$+\frac{m_2 g\sin\gamma_l}{L}\begin{bmatrix}1 & 0\\ 0 & 0\end{bmatrix}\cdot\begin{bmatrix}v_e\\ \gamma_e\end{bmatrix}. \tag{6.80}$$

In these equations of motion, the exact trajectory of the acceleration $\ddot{\gamma}_l(t)$ is not known but it is approximated by (6.79). We can also see in (6.80) that elastic structural vibrations are excited above all by high acceleration peaks. If the rigid body motion is controlled by a servomotor, then in order to avoid structural vibrations one has to take care that no acceleration peaks arise, due to sudden powering or braking for example. This finding is confirmed by engineering experience.

Furthermore, (6.80) also contains the static deflection of the beam. If in (6.80) we set all accelerations equal to zero and remove the mass point, we then directly obtain the deflection of a horizontally fixed beam loaded by its own weight

$$\begin{bmatrix}v_e\\ \gamma_e\end{bmatrix}=-\frac{m_1 g L^2}{24EJ_3}\begin{bmatrix}3L\\ 4\end{bmatrix}. \tag{6.81}$$

The exact values are always found for the static deflection from the above relations because the shape function (6.20) solves the differential equation of the elastostatics of the beam exactly.

End of Example 6.2.

Flexible multibody systems in the FFRF formulation are characterized by the separation of the rigid body motion and the elastic vibrations modelled by finite elements as shown in Sect. 6.3.1 or (6.73), respectively. While the number of the degrees of freedom f_s of rigid body motions is usually small, the number of the degrees of freedom f_e of elastic vibrations may be very large, in particular if commercial finite element codes are used. As a result the accuracy and efficiency of the simulation may strongly decrease. Therefore, it is very important to select those elastic coordinates which are interacting with the rigid bodies and are most essential for the problem under consideration. For this purpose methods of model reduction have been developed, and also applied to multibody systems.

In addition to the standard modal reduction techniques there are two methods with error control established: the moment matching by projection on Krylov subspaces, and the singular value decomposition based reduction using Gramian matrices. Both methods have been successfully used for flexible multibody system. Lehner and Eberhard [32] applied the moment matching to build reduced order models in flexible multibody dynamics. Fehr and Eberhard [21] presented the simulation process for flexible systems with Gramian reduction technique. For engineering purposes the software package Morembs[1] is available, and it is used by several companies.

6.4 Strength Calculation

While in the case of multibody systems it is only possible to estimate strength by means of the reaction forces in sections through rigid bodies, finite element systems allow a more exact calculation of strength.

The starting point of a strength computation with finite elements is Hooke's material law

$$\boldsymbol{\sigma} = \boldsymbol{H} \cdot \boldsymbol{e}. \tag{6.82}$$

The 6×1 strain vector $\boldsymbol{e}$ is thus determined by the shape functions in accordance with (2.149)

$$\boldsymbol{\sigma}_i(\boldsymbol{\rho}_i, t) = \boldsymbol{H} \cdot \boldsymbol{B}_i(\boldsymbol{\rho}_i) \cdot \boldsymbol{I}_i \cdot \boldsymbol{y}(t), \qquad i = 1(1)p. \tag{6.83}$$

We thus see that stress calculation requires knowledge of the motion $\boldsymbol{y}(t)$ of the overall system. In order to calculate strength, the stresses $\boldsymbol{\sigma}_i$ in the finite elements must be determined first. Jumps in stress arise in the transition from one element to the neighboring element. This can be reduced by selecting the right shape function. Often a solution is sought by averaging the stresses in the nodes.

In the last few decades, the finite element method has been developed and applied in structural dynamics with great success. Established program systems are currently available such as Nastran, Ansys, Permas, Abaqus, Patran, and Marc. The program systems for finite elements assume deal not only with the generation of global equations of motion but also with their solution. Input and output are usually managed by graphic user interfaces so that the user is required to have only minimal knowledge about the program system used what does usually happen. For this reason, it is often not easy for the user to make supplements or extensions to the program. It is thus advisable to design or to implement small programs oneself, at least for familiarizing with the method.

[1] www.itm.uni-stuttgart.de/research/model_reduction/model_reduction_en.php

Chapter 7
Continuous Systems

The motion of an elastic body can only be described approximately by means of the multibody system method or the finite element method. With an elaborated modelling of infinitesimal volume elements, the elastic continuum has infinitely many degrees of freedom, while its motions are determined locally by partial differential equations. First the local Cauchy equations of motion for a free continuum and for the elastic beam as a continuum with internal constraints are given, both of which must be supplemented with boundary conditions. Global equations of motion are then obtained with the eigenfunctions, which must satisfy the boundary conditions. D'Alembert's principle is applied again in this context. The global equations of motion now describe the motion of an elastic body exactly. However, this involves solving an infinite-dimensional eigenvalue problem, what is feasible only for geometrically simple bodies. For this reason, continuous systems are not as important in engineering practice as the aforementioned approximation methods. If we restrict ourselves to a finite number of eigenfunctions, such as is the case in engineering modal analysis, then continuous systems represent also an approximation method.

7.1 Local Equations of Motion

If we cut an infinitesimal volume element out of a continuum, then this element is subject to the Cauchy equations of motion (3.64), (3.65), represented with the differential operator matrix $\mathscr{V}$ in the compact form (3.67) as

$$\rho \boldsymbol{a} = \rho \boldsymbol{f} + \mathscr{V}^T \cdot \boldsymbol{\sigma}. \tag{7.1}$$

If we also take the material law (3.68) into consideration and assume small motions with respect to the inertial frame, $\boldsymbol{r}(t) = \boldsymbol{0}$, $\boldsymbol{S}(t) = \boldsymbol{0}$, then from (7.1) with (2.143) follow the local equations of motion of a free continuum

$$\rho \ddot{\boldsymbol{w}}(\boldsymbol{\rho}, t) = \mathscr{V}^T \cdot \boldsymbol{H} \cdot \mathscr{V} \cdot \boldsymbol{w}(\boldsymbol{\rho}, t) + \rho \boldsymbol{f}. \tag{7.2}$$

W. Schiehlen and P. Eberhard, *Applied Dynamics*, DOI 10.1007/978-3-319-07335-4_7,

Fig. 7.1 Boundary conditions of a continuum

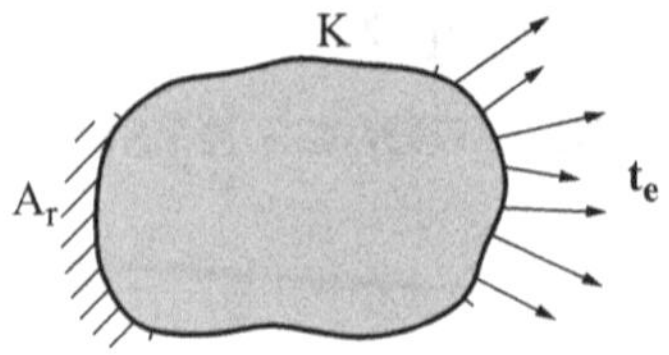

In order to solve these equations of motion, the boundary conditions on the surface of the free continuum are also required. They can either be given geometrically as constraints

$$\mathbf{w}(\boldsymbol{\rho},t) = \mathbf{w}_r(\boldsymbol{\rho},t), \qquad \boldsymbol{\rho} \in A_r, \tag{7.3}$$

and/or dynamically as applied surface forces

$$\boldsymbol{t}(\boldsymbol{\rho},t) = \boldsymbol{t}_e(\boldsymbol{\rho},t), \qquad \boldsymbol{\rho} \in A_e, \tag{7.4}$$

see Fig. 7.1. A strict solution of the equations of motion is only possible in simple cases.

However, internal constraints can be given, which often lead to useful approximations. This procedure is common, for example, in beam theory.

According to (6.12) and (6.13), the 3×1 displacement vector for a continuous beam with rigid cross-sections is

$$\mathbf{w}(\boldsymbol{\rho},t) = \underbrace{\begin{bmatrix} 1 & 0 & 0 & 0 & | & \rho_2 \frac{\partial}{\partial \rho_1} & \rho_3 \frac{\partial}{\partial \rho_1} \\ 0 & 1 & 0 & \rho_3 & | & 0 & 0 \\ 0 & 0 & 1 & -\rho_2 & | & 0 & 0 \end{bmatrix}}_{\mathscr{W}} \cdot \underbrace{\begin{bmatrix} u \\ v \\ w \\ \alpha \\ --- \\ (-v) \\ (-w) \end{bmatrix}}_{\boldsymbol{\zeta}(\rho_1,t)}, \tag{7.5}$$

where $\mathscr{W}$ is a 3×6 differential operator matrix and $\boldsymbol{\zeta}(\rho_1,t)$ a 6×1 vector of generalized functions. The six generalized functions must on the one hand be independent of each other, but on the other hand they must be linked to each other by differentiations. Therefore, the negative functions $(-v)$, $(-w)$ are carried along separately in (7.5). If we now insert (7.5) into (7.2) and apply d'Alembert's principle in the form (4.33), we then find for an infinitesimal beam element

$$\int_A \mathscr{W}^T \cdot \rho \mathscr{W} \cdot \ddot{\boldsymbol{\zeta}} dA = \int_A \mathscr{W}^T \cdot \rho \boldsymbol{f} dA + \int_A \mathscr{W}^T \cdot \mathscr{V}^T \cdot \boldsymbol{H} \cdot \mathscr{V} \cdot \mathscr{W} \cdot \boldsymbol{\zeta} dA, \tag{7.6}$$

where we should only integrate over the sectional plane A because of the infinitesimal length $d\rho_1$ of the element. If we now select a principal axis frame with the area centroid of the sectional plane as the origin, this yields decoupled equations of motion for the generalized functions u, v, w, and α.

The resultant equation for longitudinal displacement of a bar is

$$\rho A\ddot{u}(x,t) = \frac{EA}{L^2}u''(x,t) + n(x,t), \qquad x = \frac{\rho_1}{L}, \qquad u'' = \frac{\partial^2 u}{\partial x^2}, \tag{7.7}$$

with the distributed load $n(x,t)$ and the geometric boundary conditions

$$u(x_i,t) = u_r(x_i,t), \qquad x_i \in \{0,1\} \tag{7.8}$$

and/or the normal forces at the beam ends as dynamic boundary conditions

$$N_i(t) = \int_A \sigma_{11i} dA = \frac{EA}{L}u'(x_i,t), \qquad x \in \{0,1\}. \tag{7.9}$$

The coordinate ρ_1 was normalized with respect to the beam length L. The equation of motion (7.7) is a hyperbolic partial differential equation. For the torsional angle $\alpha(x,t)$, we obtain a differential equation corresponding to (7.7), where the sectional area A must be substituted with the polar area moment of inertia J_P (7.7) and Young's modulus E with the shear modulus G. By contrast, in the case of slender beams $\rho_2 \ll L$, $\rho_3 \ll L$, the deflections lead to parabolic partial differential equations. For the deflection in the 2-direction,

$$\rho A\ddot{v}(x,t) = -\frac{EJ_3}{L^4}v^{IV}(x,t) + q(x,t) \tag{7.10}$$

applies, where $q(x,t)$ represents a distributed load and the boundary conditions

$$v(x,t) = v_r(x,t), \qquad v'(x,t) = v'(x,t), \qquad x \in \{0,1\} \tag{7.11}$$

and/or bending torques and lateral forces at the beam ends

$$M(t) = \int_A \rho_2 \sigma_{11} dA = \frac{EJ_3}{L^2}v''(x,t), \tag{7.12}$$

$$Q(t) = \int_A \tau_{12} dA = \frac{EJ_3}{L^3}v'''(x,t), \qquad x \in \{0,1\}, \tag{7.13}$$

should be taken into account. Analogous relations apply for the deflection in the 3-direction.

The local equations of motion can be given in many other variants. For example, varying cross-sections and/or mass moments of inertia of the infinitesimal beam element can be taken into consideration without difficulty. Moreover, couplings between the elementary bar vibrations shown here are also possible.

7.2 Eigenfunctions of a Bar

The first step in solving partial differential equations of bar motion is the consideration of homogeneous differential equations. For example, (7.7) with $n(x,t) = 0$ yields for longitudinal displacement the homogeneous differential equation

$$\rho A\ddot{u}(x,t) = \frac{EA}{L^2}u''(x,t), \tag{7.14}$$

which must also satisfy the boundary conditions (7.8) or (7.9). With the product approach

$$u(x,t) = U(x)y(t) \tag{7.15}$$

we find, in the place of (7.14), the two ordinary differential equations

$$\ddot{y}(t) + \omega^2 y(t) = 0, \tag{7.16}$$

$$U''(t) + \beta^2 U(x) = 0, \qquad \beta^2 = \frac{\rho L^2}{E}\omega^2. \tag{7.17}$$

Since the place and time-dependent variables now appear separately, (7.15) is often also designated as a separation approach.

Equation (7.17) gives an eigenvalue problem with the general solution

$$U(x) = C_1 \sin\beta x + C_2 \cos\beta x. \tag{7.18}$$

The integration constants C_1, C_2 are determined by the boundary conditions (7.8) and/or (7.9). From the resultant characteristic equation then follow f eigenfrequencies and f eigenfunctions, $f = 1(1)\infty$, which denote the solution of the problem. With the f eigenfunctions, we can then construct the general solution.

Example 7.1 (Bar Clamped on Both Sides). The boundary conditions are exclusively determined by (7.8),

$$u(0,t) = 0, \qquad u(1,t) = 0. \tag{7.19}$$

So the characteristic equation is written

$$\sin\beta L = 0. \tag{7.20}$$

The roots of the characteristic equation lead to the eigenvalues

$$\beta_f = f\pi, \qquad f = 1(1)\infty, \tag{7.21}$$

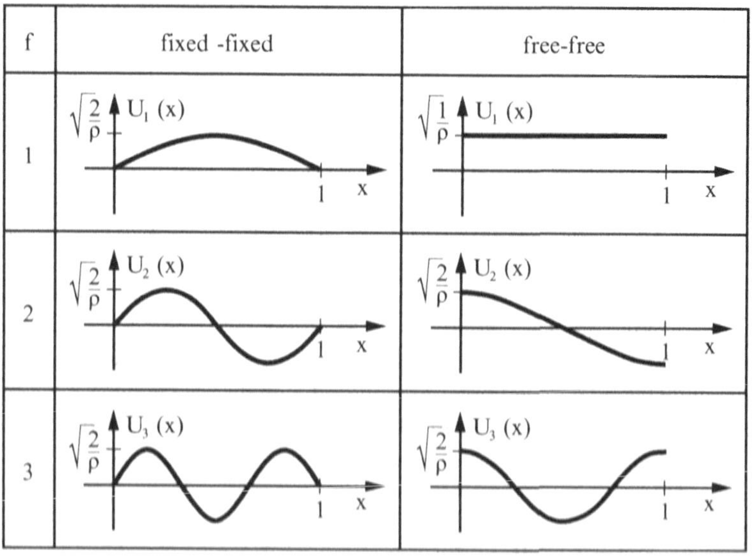

Fig. 7.2 Normalized eigenfunctions of a bar

the eigenfrequencies

$$\omega_f = f\frac{\pi}{L}\sqrt{\frac{E}{\rho}}, \qquad f = 1(1)\infty, \tag{7.22}$$

and the eigenfunctions

$$U_f(x) = C_1 \sin(f\pi x), \qquad f = 1(1)\infty. \tag{7.23}$$

Figure 7.2 shows the first three eigenfunctions. Eigenfunctions are always dependent on a scalar factor and can be normalized.

End of Example 7.1.

Example 7.2 (Free Bar). The boundary conditions are now set by (7.9), while the characteristic equation is given by

$$\beta = 0 \qquad \text{or} \qquad \sin\beta L = 0. \tag{7.24}$$

The eigenvalues are

$$\beta_f = (f-1)\pi, \qquad f = 1(1)\infty \tag{7.25}$$

and the eigenfunctions have the form

$$U_f(x) = C_2 \cos\left[(f-1)\pi x\right], \qquad f = 1(1)\infty. \tag{7.26}$$

The first eigenvalue $\beta = 0$ denotes the rigid body motion of the free bar, the higher eigenvalues describe the elastic vibrations. Figure 7.2 shows the first three eigenfunctions.

End of Example 7.2.

In the case of two time-invariant geometric boundary conditions,

$$u(x_i,t) = u_r(x_i), \qquad x_i \in \{0,1\}, \tag{7.27}$$

the eigenvalue problem always has an especially simple form, since the characteristic equation can be found directly from (7.18). But if one or two dynamic boundary conditions exist, then (7.18) must be differentiated in addition. This can be avoided if the dynamic boundary conditions are replaced with discrete volume forces acting upon the beam ends. For this purpose, the Dirac function is introduced with respect to the beam axis,

$$\delta(x-x_i) = 0 \qquad \text{for } x \neq x_i, \qquad \int_0^1 \delta(x-x_i)dx = 1. \tag{7.28}$$

We thus obtain for the equation of motion of the beam's longitudinal motion

$$\rho A\ddot{u}(x,t) = \frac{EA}{L^2}u''(x,t) + \frac{N_i(t)}{L}\delta(x-x_i) + n(x,t), \qquad x_i \in \{0,1\} \tag{7.29}$$

with the boundary conditions (7.27). The representation in (7.29) thus shifts the effects of the dynamic boundary conditions to the inhomogeneous solution component, which can be advantageous numerically. In addition, the global equations of motion can be smartly formulated with (7.29).

The procedure described above can also be applied to the deflection of a beam. For this purpose, (7.16) remains unaltered, while the eigenvalue problem assumes the form

$$U^{IV}(x) - \gamma^4 U(x) = 0, \qquad \gamma^4 = \frac{\rho A L^4}{EJ_3}\omega^2. \tag{7.30}$$

Its general solution is written

$$U(x) = C_1\cos(\gamma x) + C_2\sin(\gamma x) + C_3\cosh(\gamma x) + C_4\sinh(\gamma x), \tag{7.31}$$

where the integration constants are determined by the four boundary conditions. The reader can consult a table with the eigenvalues and eigenfunctions of varyingly fixed beams, e.g. in Holzweissig and Dresig [16] or Demeter [15].

7.3 Global Equations of Motion

A continuous system has infinitely many degrees of freedom. This is also immediately expressed in the solution of the eigenvalue problem, e.g. in (7.21). If we use the time functions $y_f(t)$, $f = 1(1)\infty$ as generalized coordinates and the eigenfunctions $U_f(t)$, $f = 1(1)\infty$ as a shape function, then the general solution for the longitudinal displacement can be expressed by

$$u(x,t) = \boldsymbol{J}(x) \cdot \boldsymbol{y}(t), \tag{7.32}$$

where

$$\boldsymbol{y}(t) = \begin{bmatrix} y_1 & y_2 & \dots & y_f \end{bmatrix}, \qquad f \to \infty, \tag{7.33}$$

represents the $f \times 1$ position vector and

$$\boldsymbol{J}(x) = \begin{bmatrix} U_1 & U_2 & \dots & U_f \end{bmatrix}, \qquad f \to \infty, \tag{7.34}$$

denotes the $f \times 1$ Jacobian matrix.

If we now insert (7.32) into the local equation of motion (7.29) and apply d'Alembert's principle (4.33) to the entire beam, the result for the normalized eigenfunctions is

$$\ddot{\boldsymbol{y}}(t) + \boldsymbol{\Omega}^2 \cdot \boldsymbol{y}(t) = \boldsymbol{q}(t). \tag{7.35}$$

In particular, normalization turns the $f \times f$ inertia matrix into a unit matrix,

$$\boldsymbol{M} = \boldsymbol{E} = \int_0^1 \boldsymbol{J}^T(x) \cdot \rho \boldsymbol{J}(x) dx, \tag{7.36}$$

which applies for all positive definite self-adjoint differential equation eigenvalue problems. We also say that the eigenfunctions are orthogonal to each other. The $f \times f$ frequency matrix

$$\boldsymbol{\Omega}^2 = \boldsymbol{diag}\{\omega_1^2 \quad \omega_2^2 \quad \dots \quad \omega_f^2\} = -\int_0^1 \boldsymbol{J}(x) \frac{E}{L^2} \cdot \boldsymbol{J}''(x) dx, \qquad f \to \infty, \tag{7.37}$$

corresponds to the positive definite stiffness matrix for the respective boundary conditions (7.27). Also, the $f \times 1$ vector of the generalized forces is given as

$$\boldsymbol{q}(t) = \frac{1}{A} \int_0^1 \boldsymbol{J}(x) n(x,t) dx + \sum_i \boldsymbol{J}(x_i) \frac{N_i(t)}{AL}, \qquad x_i \in \{0, 1\}. \tag{7.38}$$

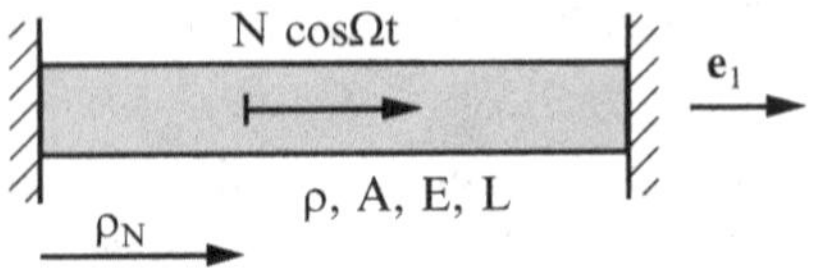

Fig. 7.3 Harmonically excited bar

The final term in (7.38) only appears if one or two dynamic boundary conditions (7.9) exist. It vanishes if only geometric boundary conditions exist, which lead to reaction forces. Otherwise, (7.38) is also valid for concentrated forces acting on any given location of the beam.

The global equations of motion (7.35) must also be supplemented with the initial conditions,

$$u(x,t_0) = \boldsymbol{J}(x) \cdot \mathbf{y}(t_0), \tag{7.39}$$

$$\dot{u}(x,t_0) = \boldsymbol{J}(x) \cdot \dot{\mathbf{y}}(t_0). \tag{7.40}$$

With the help of (7.36) we obtain

$$\mathbf{y}(t_0) = \int_0^1 \boldsymbol{J}(x)\rho u(x,t_0)dx, \tag{7.41}$$

$$\dot{\mathbf{y}}(t_0) = \int_0^1 \boldsymbol{J}(x)\rho \dot{u}(x,t_0)dx. \tag{7.42}$$

Now the global solution is completely known. The fact that the eigenfunctions (eigenforms, eigenmodes) play a decisive role in this has led to the use of the term ‘modal analysis’.

Modal analysis is also useful for beam deflection. Basically it is applicable to any continuous systems, yet successfully finding a general solution to the eigenvalue problem is rare.

Modal analysis describes one-dimensional continuous systems exactly. But this accuracy is obtained at the cost of an infinite system order. Since on the other hand it is known from vibration theory that high eigenfrequencies only contribute to a small extent to the actual motion, modal analysis is restricted in practice to a finite number of degrees of freedom. In this way, modal analysis has become yet another engineering approximation method, using the eigenfunctions as shape functions. Engineering modal analysis is thus distinguished by the fact that at least its shape functions represent exact particular solutions. Such particular solutions are especially useful for comparison with the results of multibody system and finite element methods.

Example 7.3 (Harmonically Excited Bar). A bar clamped on both sides, see Fig. 7.3, is harmonically excited by the concentrated force $N(t) = N\cos\Omega t$ at location $\rho_N = Lx_N$. According to (7.38), the following applies for the generalized force,

$$q_f(t) = \frac{N}{AL} U_f(x_N) \cos \Omega t. \tag{7.43}$$

The global equations of motion (7.35) are thus written

$$\ddot{y}_f(t) + \omega_f^2 y(t) = U_f(x_N) \frac{N}{AL} \cos \Omega t, \qquad f = 1(1)\infty \tag{7.44}$$

and for the particular solution we find

$$y_f(t) = \frac{U_f(x_N)}{\omega_f^2 - \Omega^2} \frac{N}{AL} \cos \Omega t, \qquad f = 1(1)\infty. \tag{7.45}$$

Thus, according to (7.32), the displacement of the bar is

$$u(x,t) = \sum_{f=1}^{\infty} \frac{U_f(x) U_f(x_N)}{\omega_f^2 - \Omega^2} \frac{N}{AL} \cos \Omega t. \tag{7.46}$$

The result for $\Omega \to \omega_f$ and $N \to 0$ is

$$u(x,t) = \alpha U_f(x) \cos \Omega t \tag{7.47}$$

with a constant factor α. This result corresponds to the known fact that, given a small excitation of a continuous system with the fth eigenfrequency, precisely the fth eigenform develops.

End of Example 7.3.

Just like finite element systems, continuous systems can be combined with multibody systems. As long as the additional rigid bodies follow the beam motions without friction, either the eigenvalue problem (7.17) can be reformulated or the global equations of motion (7.35) can be adequately extended. On the other hand, if multidimensional couplings and/or damping forces come into play, only the second possibility remains. This is further evidence of the limits of modal analysis.

Chapter 8
State Equations of Mechanical Systems

In the previous chapters, the equations of motion of mechanical systems were derived. These equations of motion will now be presented uniformly as state equations. The concept of state equations is especially in use in system dynamics and system theory, but it is helpful to introduce input, state, and output variables in applied dynamics as well. State variables, which include the generalized coordinates for example, can be selected in various ways. However, there are transformation laws between the selected representations. In this chapter, we will focus especially on linear systems, which can take on a normal form by means of similarity or congruence transformations.

8.1 Nonlinear State Equations

In mechanical systems, forces and motions are linked representing both input and output variables, see Fig. 8.1. If all forces are given, we can then determine the free motions as the output variables of the system. This corresponds to the direct problem of dynamics. If on the contrary all constrained motions are regarded of as input variables, then the reaction forces are the unknown output variables. Determining these is called an indirect or inverse problem of dynamics. In addition to these pure cases, there are also numerous mixed forms. In the following, only the direct problem and the differential equations of motion of the mixed problem will be discussed.

The state variables of mechanical models are given by the position variables $y_i(t)$, $i = 1(1)f$, velocity variables $z_i(t)$, $i = 1(1)g$, and force variables $w_i(t)$, $i = 1(1)h$. If we summarize all state variables with an $n \times 1$ state vector $\boldsymbol{x}(t)$, then for ordinary multibody systems, finite element systems, and continuous systems it yields

$$\boldsymbol{x}(t) = \begin{bmatrix} \mathbf{y}(t) & \dot{\mathbf{y}}(t) \end{bmatrix}, \qquad n = 2f, \tag{8.1}$$

W. Schiehlen and P. Eberhard, *Applied Dynamics*, DOI 10.1007/978-3-319-07335-4_8,

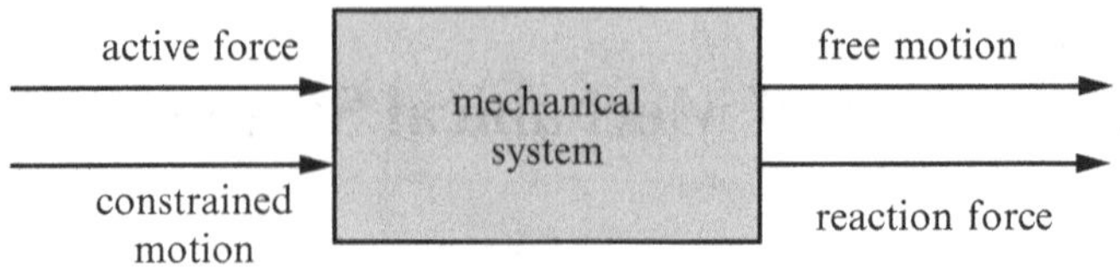

Fig. 8.1 Input and output variables of a mechanical system

while for general multibody systems

$$\boldsymbol{x}(t) = \left[\ \boldsymbol{y}(t)\ \ \boldsymbol{z}(t)\ \ \boldsymbol{w}(t)\ \right], \qquad n = f + g + h \tag{8.2}$$

applies. The input variables are determined by the available control forces and other forces, generally designated as disturbances. The r input variables are merged into an $r \times 1$ input vector

$$\boldsymbol{u} = \boldsymbol{u}(t), \qquad r \leq n. \tag{8.3}$$

In applied dynamics, the output variables often correspond to individual state variables, but linear combinations of input and output variables are also possible.

The global equations of motion (5.28), (5.35), (5.74), (6.40), and (7.35) can each be represented with (8.1)–(8.3) by the nonlinear state equations

$$\dot{\boldsymbol{x}}(t) = \boldsymbol{a}(\boldsymbol{x}, \boldsymbol{u}, t), \qquad \boldsymbol{x}(t_0) = \boldsymbol{x}_0, \tag{8.4}$$

where $\boldsymbol{a}$ denotes an $n \times 1$ vector function and $\boldsymbol{x}_0$ the $n \times 1$ vector of the initial conditions.

8.2 Linear State Equations

By linearizing, we obtain from the nonlinear state equations (8.4), insofar as it is physically permissible, the linear state equations, which are extremely important in engineering practice

$$\dot{\boldsymbol{x}}(t) = \boldsymbol{A}(t) \cdot \boldsymbol{x}(t) + \boldsymbol{B}(t) \cdot \boldsymbol{u}(t), \qquad \boldsymbol{x}(t_0) = \boldsymbol{x}_0. \tag{8.5}$$

Here, $\boldsymbol{A}(t)$ is the $n \times n$ system matrix and $\boldsymbol{B}(t)$ the $n \times r$ input matrix. The state equations (8.5) denote a time-variant system, while for constant matrices $\boldsymbol{A}$ and $\boldsymbol{B}$ a time-invariant system is given.

The matrices have a characteristic structure for ordinary linear multibody systems. Equation (5.59) yields with (8.1) and $\boldsymbol{u}(t) = \boldsymbol{h}(t)$

$$\boldsymbol{A} = \left[\begin{array}{c|c} \boldsymbol{0} & \boldsymbol{E} \\ \hline -\boldsymbol{M}^{-1} \cdot \boldsymbol{Q} & -\boldsymbol{M}^{-1} \cdot \boldsymbol{P} \end{array} \right], \qquad \boldsymbol{B} = \left[\begin{array}{c} \boldsymbol{0} \\ \hline \boldsymbol{M}^{-1} \end{array} \right], \tag{8.6}$$

which also apply for finite element systems and continuous systems. In order to evaluate the matrices $\boldsymbol{A}$ and $\boldsymbol{B}$, we always need the inverse $\boldsymbol{M}^{-1}$ of the inertia matrix in the case of mechanical systems. Since the inertia matrix of linear systems is positive definite, its inverse always exists.

8.3 Transformation of Linear Equations

The choice of state variables influences the matrices of the state equations (8.5). Since the course of a mechanical motion cannot depend on the choice of state variables, there must be a mathematical relationship between different descriptions. This relationship is arranged by transformation laws.

For linear, time-invariant systems, the relation between two different $n \times 1$ state vectors $\boldsymbol{x}(t)$ and $\hat{\boldsymbol{x}}(t)$ is given by a constant and regular $n \times n$ transformation matrix $\boldsymbol{T}$,

$$\boldsymbol{x}(t) = \boldsymbol{T} \cdot \hat{\boldsymbol{x}}(t). \tag{8.7}$$

Inserting (8.7) into the state equations (8.5), we obtain

$$\boldsymbol{T} \cdot \dot{\hat{\boldsymbol{x}}}(t) = \boldsymbol{A} \cdot \boldsymbol{T} \cdot \hat{\boldsymbol{x}}(t) + \boldsymbol{B} \cdot \boldsymbol{u}(t) \tag{8.8}$$

or

$$\dot{\hat{\boldsymbol{x}}}(t) = \hat{\boldsymbol{A}} \cdot \hat{\boldsymbol{x}}(t) + \hat{\boldsymbol{B}} \cdot \boldsymbol{u}(t) \tag{8.9}$$

with the transformation law for the system matrix

$$\hat{\boldsymbol{A}} = \boldsymbol{T}^{-1} \cdot \boldsymbol{A} \cdot \boldsymbol{T} \tag{8.10}$$

and the input matrix

$$\hat{\boldsymbol{B}} = \boldsymbol{T}^{-1} \cdot \boldsymbol{B}. \tag{8.11}$$

The transformation law (8.10) describes a similarity transformation. This generally destroys the particular structure of the matrices (8.6). However, there are special transformation matrices that do not alter the structure of the matrices (8.6). Such a transformation matrix is given by

$$\boldsymbol{T} = \left[\begin{array}{c|c} \boldsymbol{U} & \boldsymbol{0} \\ \hline \boldsymbol{0} & \boldsymbol{U} \end{array}\right], \tag{8.12}$$

where $\boldsymbol{U}$ is a constant, regular $f \times f$ matrix.

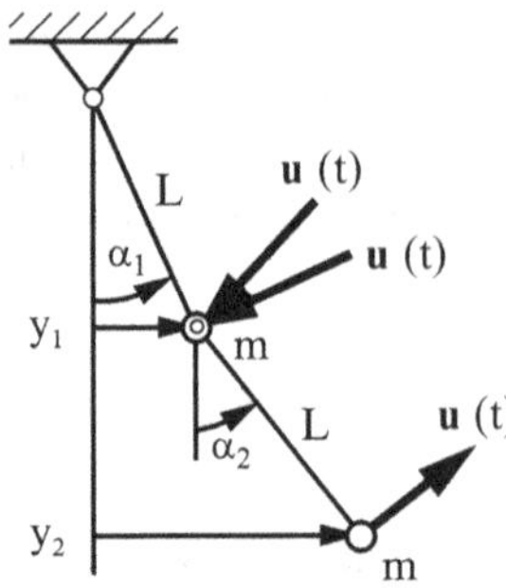

Fig. 8.2 Double pendulum with force control

Just like the state variables, the position variables can also be selected arbitrarily. Then the following relation exists between two different position vectors $\mathbf{y}(t)$ and $\hat{\mathbf{y}}(t)$,

$$\mathbf{y}(t) = \boldsymbol{U} \cdot \hat{\mathbf{y}}(t). \tag{8.13}$$

If we insert (8.13) into (5.60), we then find for linear, time-invariant systems

$$\boldsymbol{M} \cdot \boldsymbol{U} \cdot \ddot{\hat{\mathbf{y}}}(t) + (\boldsymbol{D}+\boldsymbol{G}) \cdot \boldsymbol{U} \cdot \dot{\hat{\mathbf{y}}}(t) + (\boldsymbol{K}+\boldsymbol{N}) \cdot \boldsymbol{U} \cdot \hat{\mathbf{y}}(t) = \boldsymbol{h}(t) \tag{8.14}$$

where $\boldsymbol{M} \cdot \boldsymbol{U}$ represents an unsymmetrical matrix. If the original symmetry of the inertia matrix is again required, then (8.14) must be multiplied from the left with $\boldsymbol{U}^T$. This results in

$$\hat{\boldsymbol{M}} \cdot \ddot{\hat{\mathbf{y}}}(t) + (\hat{\boldsymbol{D}}+\hat{\boldsymbol{G}}) \cdot \dot{\hat{\mathbf{y}}}(t) + (\hat{\boldsymbol{K}}+\hat{\boldsymbol{N}}) \cdot \hat{\mathbf{y}}(t) = \hat{\boldsymbol{h}}(t), \tag{8.15}$$

from which we find the transformation law for all matrices

$$\hat{\boldsymbol{M}} = \boldsymbol{U}^T \cdot \boldsymbol{M} \cdot \boldsymbol{U} \tag{8.16}$$

and for the excitation vector

$$\hat{\boldsymbol{h}}(t) = \boldsymbol{U}^T \cdot \boldsymbol{h}(t). \tag{8.17}$$

The transformation law (8.16) is called a congruence transformation.

When setting up the equations of motion according to d'Alembert's principle, congruence transformations appear again and again, whereby not only quadratic transformation matrices, e.g. in (5.29), but also rectangular transformation matrices can be found, e.g. in (5.36), (6.41), or (7.36). D'Alembert's principle can therefore itself also be understood as a congruence transformation in the wider sense. Its characteristics is the preservation of the symmetry of the inertia matrix.

Example 8.1 (Double Pendulum). The linearized equations of motion of an actuated double pendulum, see Fig. 8.2, are for small angles

$$mL^2 \begin{bmatrix} 2 & 1 \\ 1 & 1 \end{bmatrix} \cdot \begin{bmatrix} \ddot{\alpha}_1 \\ \ddot{\alpha}_2 \end{bmatrix} + mgL \begin{bmatrix} 2 & 0 \\ 0 & 1 \end{bmatrix} \cdot \begin{bmatrix} \alpha_1 \\ \alpha_2 \end{bmatrix} = \begin{bmatrix} -L \\ L \end{bmatrix} u(t). \tag{8.18}$$

The actuation comes from three equally large control forces $u(t)$, which are perpendicular to both bars of the double pendulum.

If the horizontal excursions y_1 and y_2 are used instead of the angles as generalized coordinates, then for small angles we obtain the transformation matrix

$$\boldsymbol{U} = \frac{1}{L}\begin{bmatrix} 1 & 0 \\ -1 & 1 \end{bmatrix}. \tag{8.19}$$

According to (8.15), (8.16), from this follow the congruently transformed equations of motion

$$m\begin{bmatrix} 1 & 0 \\ 0 & 1 \end{bmatrix}\cdot\begin{bmatrix} \ddot{y}_1 \\ \ddot{y}_2 \end{bmatrix} + \frac{mg}{L}\begin{bmatrix} 3 & -1 \\ -1 & 1 \end{bmatrix}\cdot\begin{bmatrix} y_1 \\ y_2 \end{bmatrix} = \begin{bmatrix} -2 \\ 1 \end{bmatrix}u(t). \tag{8.20}$$

We can see that the inertia forces are coupled in (8.18), while in (8.20) the weights are coupled. The type of coupling thus depends on the choice of coordinates. It is therefore impossible to obtain information about the mechanical behavior of a linear system from the type of coupling. On the other hand, the explanation of the terms can be simplified by the right choice of coordinates. For example, the generalized forces in (8.20) directly correspond to the applied mechanical forces, see Fig. 8.2.

Also, the equations of motion (8.18) must still be transferred into the state equations. According to (8.1), the 4×1 state vector is written

$$\boldsymbol{x}(t) = \begin{bmatrix} \alpha_1 & \alpha_2 & \dot{\alpha}_1 & \dot{\alpha}_2 \end{bmatrix} \tag{8.21}$$

and the matrices of the state equations have the form

$$\boldsymbol{A} = \left[\begin{array}{cc|cc} 0 & 0 & 1 & 0 \\ 0 & 0 & 0 & 1 \\ \hline -2\frac{g}{L} & \frac{g}{L} & 0 & 0 \\ 2\frac{g}{L} & -2\frac{g}{L} & 0 & 0 \end{array}\right], \quad \boldsymbol{B} = \left[\begin{array}{c} 0 \\ 0 \\ \hline -\frac{2}{mL} \\ \frac{3}{mL} \end{array}\right]. \tag{8.22}$$

We can see that the submatrices of the system matrix $\boldsymbol{A}$ generally lose their symmetry properties.

End of Example 8.1.

8.4 Normal Forms

The choice of state variables affects the coefficient matrices of the state equations. Thus the question rises whether state variables exist that give the system matrix a form that is as simple as possible. This question will be pursued for the homogeneous system

$$\dot{\boldsymbol{x}}(t) = \boldsymbol{A}\cdot\boldsymbol{x}(t), \tag{8.23}$$

the solution of which should also satisfy for the decoupled system

$$\dot{\boldsymbol{x}}(t) = \lambda \boldsymbol{E} \cdot \boldsymbol{x}(t). \tag{8.24}$$

First, (8.24) yields the particular solution

$$\boldsymbol{x}(t) = e^{\lambda t}\tilde{\boldsymbol{x}} \tag{8.25}$$

with the constant $n \times 1$ vector $\tilde{\boldsymbol{x}}$ of the initial conditions. Inserted into (8.23), we then obtain the associated eigenvalue problem

$$(\lambda \boldsymbol{E} - \boldsymbol{A}) \cdot \tilde{\boldsymbol{x}} = \boldsymbol{0}. \tag{8.26}$$

The eigenvalue problem thus represents a homogeneous algebraic equation system, which only has a nontrivial solution if the characteristic matrix $(\lambda \boldsymbol{E} - \boldsymbol{A})$ is singular.

The requirement

$$\det(\lambda \boldsymbol{E} - \boldsymbol{A}) = 0 \tag{8.27}$$

provides the eigenvalues λ_i, $i = 1(1)n$, which should all be different in the following. According to (8.26), to each eigenvalue λ_i belongs one eigenvector $\tilde{\boldsymbol{x}}_i, i = 1(1)n$. If these eigenvectors are summarized in an $n \times n$ modal matrix

$$\boldsymbol{X} = \left[\ \tilde{\boldsymbol{x}}_1 \ \mid \ \tilde{\boldsymbol{x}}_2 \ \mid \ \dots \ \mid \ \tilde{\boldsymbol{x}}_n \ \right], \tag{8.28}$$

we then have a transformation matrix that converts the state equations (8.23) into their normal form

$$\dot{\hat{\boldsymbol{x}}}(t) = \boldsymbol{\Lambda} \cdot \hat{\boldsymbol{x}}(t), \tag{8.29}$$

where

$$\boldsymbol{\Lambda} = \boldsymbol{diag}\{\lambda_i\}, \qquad i = 1(1)n, \tag{8.30}$$

is the $n \times n$ diagonal matrix of the eigenvalues. The marked state variables $\hat{\boldsymbol{x}}_i(t)$, $i = 1(1)n$ are called principal or normal coordinates. They are nice mathematically because of the decoupling, but they generally lack an intuitive mechanical meaning due to (8.28). The eigenvectors $\tilde{\boldsymbol{x}}_i$ represent eigenforms or vibration forms, which occur, for example, if the system is operating under resonance conditions.

In the case of multiple eigenvalues, if may be necessary to substitute the diagonal matrix by a Jordan matrix. For the definition and computation of the Jordan matrix, see the relevant literature, e.g., Müller and Schiehlen [37].

A suitable choice of position variables can further simplify the system matrices of the equations of motion. The solution of the conservative system

$$\boldsymbol{M} \cdot \ddot{\boldsymbol{y}}(t) + \boldsymbol{K} \cdot \boldsymbol{y}(t) = \boldsymbol{0} \tag{8.31}$$

should also satisfy the simplified system

$$\ddot{\boldsymbol{y}}(t) + \omega^2 \boldsymbol{E} \cdot \boldsymbol{y}(t) = \boldsymbol{0}. \tag{8.32}$$

With the particular solution

$$\boldsymbol{y}(t) = \sin \omega t \, \tilde{\boldsymbol{y}} \tag{8.33}$$

of (8.32), (8.31) yields the eigenvalue problem

$$(-\boldsymbol{M}\omega^2 + \boldsymbol{K}) \cdot \tilde{\boldsymbol{y}} = \boldsymbol{0}. \tag{8.34}$$

The characteristic equation

$$\det(-\boldsymbol{M}\omega^2 + \boldsymbol{K}) = 0 \tag{8.35}$$

now supplies the eigenfrequencies $\omega_j, j = 1(1)f$, which should again all be assumed to be different. With the normalized $f \times f$ modal matrix

$$\hat{\boldsymbol{Y}} = \boldsymbol{Y} \cdot (\boldsymbol{Y}^T \cdot \boldsymbol{M} \cdot \boldsymbol{Y})^{-\frac{1}{2}}, \qquad \boldsymbol{Y} = \left[\; \tilde{\boldsymbol{y}}_1 \; | \; \tilde{\boldsymbol{y}}_2 \; | \; \dots \; | \; \tilde{\boldsymbol{y}}_f \; \right], \tag{8.36}$$

the equations of motion (8.31) are transformed in accordance with (8.16) to their normal form

$$\ddot{\hat{\boldsymbol{y}}}(t) + \boldsymbol{\Omega}^2 \cdot \hat{\boldsymbol{y}}(t) = \boldsymbol{0}, \tag{8.37}$$

and we thereby find the $f \times f$ diagonal matrix

$$\boldsymbol{\Omega}^2 = \boldsymbol{\Omega} \cdot \boldsymbol{\Omega} = \boldsymbol{diag}\{\omega_j^2\}, \qquad j = 1(1)f. \tag{8.38}$$

If we now apply the modal transformation, (8.16) with (8.36), to the equations of motion of ordinary multibody systems (5.60), we then obtain

$$\ddot{\hat{\boldsymbol{y}}}(t) + (\hat{\boldsymbol{D}} + \hat{\boldsymbol{G}}) \cdot \dot{\hat{\boldsymbol{y}}}(t) + \boldsymbol{\Omega}^2 \cdot \hat{\boldsymbol{y}}(t) + \hat{\boldsymbol{N}} \cdot \hat{\boldsymbol{y}}(t) = \hat{\boldsymbol{h}}(t). \tag{8.39}$$

It should be noted that the matrices $\hat{\boldsymbol{D}}$, $\hat{\boldsymbol{G}}$, $\hat{\boldsymbol{N}}$ generally do not exhibit a diagonal form. This means that the congruence transformation only allows the decoupling of conservative, non-gyroscopic systems. This issue was already addressed in Chap. 7 in the discussion of the modal analysis of continuous systems.

The generalized coordinates $\hat{y}_j(t), j = 1(1)f$, too are called principal or normal coordinates. These coordinates no longer describe the motion of single points or bodies of a mechanical system, but rather represent a linear combination of mechanical coordinates. This means that single normal or principal coordinates also cannot be determined by measuring the motion of single points or bodies.

On the other hand, the vector $\hat{\boldsymbol{y}}(t)$ of the normal coordinates and the vector $\hat{\boldsymbol{h}}(t)$ of the generalized forces are closely linked. In particular, the generalized forces $\hat{h}_j, j = 1(1)f$, correspond not to single mechanical forces and torques but are linear combinations thereof.

This once again clarifies a difference between the finite element method and the method of continuous systems. In finite element systems, usually the mechanical absolute or relative coordinates are selected as position variables, and the associated generalized forces are mechanical forces and torques. In contrast, the equations of motion of continuous systems are often based on the normal and principal coordinates, while the generalized forces are infinite-dimensional linear combinations of mechanical forces. Therefore, continuous systems are not as suitable for the development of clearly structured computation methods and computer-compatible program systems.

Example 8.2 (Double Pendulum). The normal form of the equations of motion (8.20) of the double pendulum, see Fig. 8.2, can be obtained with the normalized modal matrix

$$\hat{\boldsymbol{Y}} = \frac{1}{\sqrt{m}} \begin{bmatrix} \frac{1}{\sqrt{4+2\sqrt{2}}} & \frac{1}{\sqrt{4-2\sqrt{2}}} \\ \frac{1+\sqrt{2}}{\sqrt{4+2\sqrt{2}}} & \frac{1-\sqrt{2}}{\sqrt{4-2\sqrt{2}}} \end{bmatrix} \tag{8.40}$$

from (8.20). The congruence transformation directly yields

$$\begin{bmatrix} \ddot{\hat{y}}_1 \\ \ddot{\hat{y}}_2 \end{bmatrix} + \frac{g}{L} \begin{bmatrix} 2-\sqrt{2} & 0 \\ 0 & 2+\sqrt{2} \end{bmatrix} \cdot \begin{bmatrix} \hat{y}_1 \\ \hat{y}_2 \end{bmatrix} = \begin{bmatrix} \frac{-1+\sqrt{2}}{\sqrt{4+2\sqrt{2}}} \\ \frac{-1-\sqrt{2}}{\sqrt{4-2\sqrt{2}}} \end{bmatrix} \frac{1}{\sqrt{m}} u(t). \tag{8.41}$$

The normal form (8.41) can also be obtained from (8.18). The eigenvectors, and with them the modal matrix $\hat{\boldsymbol{Y}}$, depend on the selection of position variables. On the other hand, the eigenfrequencies remain invariant after congruence transformations. The generalized forces $\hat{\boldsymbol{h}}(t)$ in (8.41) no longer have anything to do with the mechanical forces and torques $u(t)$ in (8.20). Even their dimension was changed by the modal transformation.

End of Example 8.2.

Nonlinear multibody systems can also be converted into a normal form with respect to the inertia matrix, as can be shown in Sect. 5.3.2. However, since the transformation matrix depends in the nonlinear case on the position variables, the result is a position-dependent eigenvalue problem. Because of the resulting computational complexity, the concept of the normal form has not become important for nonlinear engineering systems.

Chapter 9
Numerical Methods

In applied dynamics, large motions lead to coupled, nonlinear systems of ordinary differential equations, small motions result in linear systems of differential equations, and finally reaction forces are determined by algebraic systems of equations. In order to solve these problems, numerical mathematics offers many established methods which are available to applied dynamics. Yet this was not always the case. For example, the classic work by Biezeno and Grammel [10] still contains an extensive chapter about solution methods for eigenvalue and boundary value problems. In particular, numerical mathematics has again and again gotten strong stimuli for its own further development from applied dynamics.

Many methods of numerical mathematics are available today in a highly user-friendly format, e.g. in the software Matlab. Matlab programs can be used to great benefit for the solution of numerical problems in applied dynamics. It is nonetheless recommendable that one become familiar with the basic ideas of the solution methods involved.

In this chapter, we will briefly discuss some of the important numerical methods used in applied dynamics. These include methods of time integration of ordinary systems of differential equations and linear algebra for time-invariant systems. Finally, the mechanical models that have been introduced will be compared with the help of some numerical results.

9.1 Integration of Nonlinear Differential Equations

For numerical integration, first the time function of the input variables (8.3) are inserted into the state equations (8.4), giving the $n \times 1$ vector differential equation

$$\dot{\boldsymbol{x}}(t) = \boldsymbol{a}(\boldsymbol{x},t), \qquad \boldsymbol{x}(t_0) = \boldsymbol{x}_0. \tag{9.1}$$

Several numerical integration methods are available for this, which have been described extensively e.g. by Grigorieff [24] or Butcher [14]. The solution of (9.1)

W. Schiehlen and P. Eberhard, *Applied Dynamics*, DOI 10.1007/978-3-319-07335-4_9,

depends on the initial state $\boldsymbol{x}(t_0)$, which is why we also call it an initial value problem. The various methods used for solving initial value problems can be subdivided into one-step methods, multistep methods, and extrapolation methods. These procedures, each of which have many variants, will now be briefly introduced. The best-known of the one-step methods are the Runge-Kutta methods. In it, (9.1) is solved step-by-step proceeding from the initial state. With the one-step methods, the state

$$\boldsymbol{x}(t_{k+1}) = \boldsymbol{x}(t_k + h) \tag{9.2}$$

is computed at the discrete time instants $t_k = t_0 + kh, k = 1,2,3,\ldots$ and time increment h, using an approximate formula. The latter is, e.g. for the well-known Runge-Kutta method of the fourth order

$$\boldsymbol{x}(t_{k+1}) \approx \boldsymbol{x}(t_k) + \frac{h}{6}[\boldsymbol{\Delta x}_k^{(1)} + 2\boldsymbol{\Delta x}_k^{(2)} + 2\boldsymbol{\Delta x}_k^{(3)} + \boldsymbol{\Delta x}_k^{(4)}] \tag{9.3}$$

with the $n \times 1$ auxiliary vectors

$$\boldsymbol{\Delta x}_k^{(1)} = \boldsymbol{a}(\boldsymbol{x}(t_k), t_k), \tag{9.4}$$

$$\boldsymbol{\Delta x}_k^{(2)} = \boldsymbol{a}(\boldsymbol{x}(t_k) + \frac{1}{2}\boldsymbol{\Delta x}_k^{(1)}, t_k + \frac{1}{2}h), \tag{9.5}$$

$$\boldsymbol{\Delta x}_k^{(3)} = \boldsymbol{a}(\boldsymbol{x}(t_k) + \frac{1}{2}\boldsymbol{\Delta x}_k^{(2)}, t_k + \frac{1}{2}h), \tag{9.6}$$

$$\boldsymbol{\Delta x}_k^{(4)} = \boldsymbol{a}(\boldsymbol{x}(t_k) + \boldsymbol{\Delta x}_k^{(3)}, t_k + h). \tag{9.7}$$

In order to better understand these formulae, (9.3) and (9.4) can be applied to a homogeneous time-invariant system

$$\dot{\boldsymbol{x}}(t) = \boldsymbol{A} \cdot \boldsymbol{x}(t). \tag{9.8}$$

We then find

$$\boldsymbol{x}(t_{k+1}) \approx \boldsymbol{\phi}(h) \cdot \boldsymbol{x}(t_k) \tag{9.9}$$

with

$$\boldsymbol{\phi}(h) = \boldsymbol{E} + \boldsymbol{A}h + \frac{1}{2}\boldsymbol{A} \cdot \boldsymbol{A}h^2 + \frac{1}{6}\boldsymbol{A} \cdot \boldsymbol{A} \cdot \boldsymbol{A}h^3 + \frac{1}{24}\boldsymbol{A} \cdot \boldsymbol{A} \cdot \boldsymbol{A} \cdot \boldsymbol{A}h^4. \tag{9.10}$$

The error of one integration step is in the order of h^5.

The most important variable in a one-step method is the increment h. It has a direct influence on the integration error. In particular, we can improve the accuracy of the result by computing with smaller increments, although rounding errors set limits to this. The execution of such control computations can be automated by an increment control. With such automated methods, it is no longer the increment but a tolerable discretization error that is given.

The order of the method can also be increased. Raising the order of the method reduces error as well, but computational complexity is increased due to the additional terms in the approximate formulae. It is possible in one method to control not only the increment but also the order during the computation. An example of this is the Runge-Kutta-Fehlberg method of the fifth/sixth order with increment control.

The advantages of the one-step methods are that they are simple to program and that the calculation starts directly from the initial state. Their disadvantage is that information about the solution path already calculated cannot be exploited in the further calculation. This leads to a large amount of function evaluations of (9.1). Furthermore, increment control is essential for exact results. Nevertheless, experience has shown that one can already obtain good and safe results in applied dynamics with the aforementioned Runge-Kutta-Fehlberg method of the fifth/sixth order.

Among multistep methods, predictor-corrector methods are especially worthy of mention. These also make use of information from the solution path that has already been calculated, too. It is immediately obvious that this increases the programming effort considerably. In addition, instabilities can arise with multistep methods just like with one-step methods, and reducing the increments does not necessarily lead to improved accuracy in every multistep method. The start of the computation is a further problem, since additional information beyond the initial state is not available at the starting time. Such disadvantages of multistep methods can however be largely minimized as e.g. the Shampine-Gordon method shows. We are then left with the advantage of short calculation times with high levels of accuracy.

The method of Shampine and Gordon [58] is a multistep method with order and increment control. It is self-starting and is characterized by very few function evaluations of (9.1). The method discovers and controls discontinuities, it also controls rounding errors given high accuracy requirements, it reports excessive accuracy demands, and it also controls to a certain extent larger frequency differences in differential equation systems. The Shampine-Gordon method is comfortable and supplies an arbitrary number of intermediate values for the graphic representation of solution curves independently of the internal increment – it has been grounded, documented, and tested to an outstanding extent. The Shampine-Gordon method is highly suitable for problems in applied dynamics and has proven itself excellently.

Extrapolation methods work with a sequence of increments that permit an extrapolation to a limit. For this, polynomial or rational functions are used in conjunction with recursion formulae, which again results in a relatively large number of function evaluations of (9.1). The advantages of extrapolation methods are their high levels of accuracy with comparatively large increments. If this property is not of primary concern, then the higher computation complexity is not worthwhile. One widespread extrapolation method is the Gragg-Bulirsch-Stoer method. It has proved extraordinarily useful for applications in celestial mechanics, where large increments are decisive. For applied dynamics, extrapolation methods are usually less suitable.

9.2 Linear Algebra of Time-Invariant Systems

While the nonlinear state equations (9.1) can practically only be solved as an initial value problem, the linear, time-invariant state equations

$$\dot{\boldsymbol{x}}(t) = \boldsymbol{A} \cdot \boldsymbol{x}(t) + \boldsymbol{b}(t), \qquad \boldsymbol{x}(t_0) = \boldsymbol{x}_0 \tag{9.11}$$

can also be investigated in a purely algebraic manner. The corresponding solution methods, also used for linear vibration systems, are shown e.g. in Müller and Schiehlen [37]. Here we will only introduce a few basic ideas.

The general solution of (9.11) is

$$\boldsymbol{x}(t) = \boldsymbol{\phi}(t) \cdot \boldsymbol{x}_0 + \int_0^t \boldsymbol{\phi}(t-\tau) \cdot \boldsymbol{b}(\tau) d\tau, \tag{9.12}$$

where $\boldsymbol{\phi}(t)$ is the $n \times n$ fundamental matrix, which can be determined by the eigenvalue problem. With (8.28) and (8.30)

$$\boldsymbol{\phi}(t) = \boldsymbol{X} \cdot e^{\boldsymbol{\Lambda} t} \cdot \boldsymbol{X}^{-1} \tag{9.13}$$

applies, i.e. the fundamental matrix depends solely on the eigenvalues and the modal matrix and can be determined without numerical integration.

Equation (9.12) also contains the integral of the particular solution. This integral can be solved analytically if the $n \times 1$ excitation vector $\boldsymbol{b}(t)$ is either temporally limited, periodical, or stochastic. The excitation vector $\boldsymbol{b}(t)$ is defined according to (8.5) by the input matrix $\boldsymbol{B}(t)$ and the input vector $\boldsymbol{u}(t)$.

A temporally limited excitation can be approximated by a polynomial of the first degree,

$$\boldsymbol{b}(t) = \sum_{k=0}^{l} \boldsymbol{b}_k \{t - t_A, t_E\}^k, \tag{9.14}$$

where the $n \times 1$ coefficient vectors $\boldsymbol{b}_k$ and the scalar window function

$$\{t - t_A, t_E\}^k = \begin{cases} 0 & \text{for } t \leq t_A, \\ (t-t_A)^k & \text{for } t_A < t < t_E, \\ 0 & \text{for } t_E \leq t \end{cases} \tag{9.15}$$

are used. Then the general solution is written

$$\begin{aligned} \boldsymbol{x}(t) = \boldsymbol{\phi}(t) \cdot \Bigg[& \boldsymbol{x}_0 + \boldsymbol{\phi}(-t_A) \cdot \boldsymbol{f}_0 \{t - t_A, \infty\}^0 \\ & - \boldsymbol{\phi}(-t_E) \cdot \sum_{k=0}^{l} \boldsymbol{f}_k (t_E - t_A)^k \{t - t_E, \infty\}^0 \Bigg] - \sum_{k=0}^{l} \boldsymbol{f}_k \{t_E - t_A, t_E\}^k \end{aligned} \tag{9.16}$$

with the abbreviation

$$\boldsymbol{f}_k = \sum_{m=k}^{l} \frac{m!}{k!} \boldsymbol{A}^{(k-m-1)} \cdot \boldsymbol{b}_m. \tag{9.17}$$

The answer (9.16) is distinguished by three characteristic time periods. In the first interval $0 < t < t_A$, a free vibration with the initial condition $\boldsymbol{x}_0$ is taking place. In the second interval $t_A < t < t_E$, a free vibration and a forced vibration overlap, while in the third interval $t_E < t$ a free vibration appears with an altered initial condition.

A periodic excitation can be approximated by a Fourier series of 1st order

$$\boldsymbol{b}(t) = \sum_{k=1}^{l} (\boldsymbol{b}_k^{(1)} \cos k\Omega t + \boldsymbol{b}_k^{(2)} \sin k\Omega t) \tag{9.18}$$

with the $n \times 1$ vectors $\boldsymbol{b}_k^{(1)}, \boldsymbol{b}_k^{(2)}$ of the Fourier coefficients. The general solution now has the form

$$\boldsymbol{x}(t) = \boldsymbol{\phi}(t) \cdot \left[\boldsymbol{x}_0 - \sum_{k=1}^{l} \boldsymbol{g}_k^{(1)} \right] + \sum_{k=1}^{l} (\boldsymbol{g}_k^{(1)} \cos k\Omega t + \boldsymbol{g}_k^{2} \sin k\Omega t) \tag{9.19}$$

with the complex $n \times 1$ frequency response vector

$$\boldsymbol{g}_k^{(1)} - i\boldsymbol{g}_k^{(2)} = (ik\Omega \boldsymbol{E} - \boldsymbol{A})^{-1} \cdot (\boldsymbol{b}_k^{(1)} - i\boldsymbol{b}_k^{(2)}). \tag{9.20}$$

The answer (9.19) represents the overlapping of a free vibration that describes the transient effect and a forced vibration.

For the stochastic excitation of asymptotically stable systems, we assume a Gaussian $n \times 1$ vector process

$$\boldsymbol{b}(t, \tau) \sim (\boldsymbol{m}(t), \boldsymbol{N}_b(t, \tau)) \tag{9.21}$$

with the properties of white noise. Then, it yields for the $n \times 1$ mean vector $\boldsymbol{m}_b(t) = \boldsymbol{0}$ and $\boldsymbol{N}_b(t, \tau) = \boldsymbol{Q}\delta(t - \tau)$ for the $n \times n$ correlation matrix. Here, $\boldsymbol{Q}$ is the $n \times n$ intensity matrix of the white noise, and δ describes the Dirac function introduced by (7.28). If we also take into consideration that in stochastic systems the initial state is also described by a Gaussian $n \times 1$ random vector

$$\boldsymbol{x}_0 \sim (\boldsymbol{m}_0, \boldsymbol{P}_0) \tag{9.22}$$

with the $n \times 1$ mean vector $\boldsymbol{m}_0$ and the $n \times n$ covariance matrix $\boldsymbol{P}_0$, we then also find the general solution as the resulting Gaussian vector process

$$\boldsymbol{x}(t,\tau) \sim (\boldsymbol{m}_x(t), \boldsymbol{N}_x(t,\tau)). \tag{9.23}$$

It is extremely remarkable that, for the $n \times 1$ mean vector $\boldsymbol{m}_x(t)$ and the $n \times n$ covariance matrix $\boldsymbol{P}_x(t)$ resulting from the $n \times n$ correlation matrix $\boldsymbol{N}_x(t,\tau)$, the deterministic solutions

$$\boldsymbol{m}_x(t) = \boldsymbol{\phi}(t) \cdot \boldsymbol{m}_0, \tag{9.24}$$

$$\boldsymbol{P}_x(t) = \boldsymbol{\phi}(t) \cdot (\boldsymbol{P}_0 - \boldsymbol{P}) \cdot \boldsymbol{\phi}^T(t) + \boldsymbol{P} \tag{9.25}$$

could be found, which can be traced back to the fundamental matrix (9.13) and the algebraic Ljapunov matrix equation

$$\boldsymbol{A} \cdot \boldsymbol{P} + \boldsymbol{P} \cdot \boldsymbol{A}^T + \boldsymbol{Q} = \boldsymbol{0}. \tag{9.26}$$

The answer (9.24) describes a pure transient vibration to the vanishing mean, while the covariance matrix (9.25) is a transient vibration to a steady value. For further details, see Müller and Schiehlen [37].

The aforementioned restriction to excitation via white noise is not very serious because colored noise processes can be generated using "linear form filters". For this purpose, the system order of the state equation (9.11) may be slightly raised.

It has thus been demonstrated that the applied dynamics of time-invariant systems can be reduced to purely algebraic problems. The question remains of which methods of linear algebra can be used most effectively.

The numerical solution of the eigenvalue problem takes advantage on a large scale of repeated similarity transformations. It is recommended that one first balances the given system matrix $\boldsymbol{A}$ in order to improve its condition. Then a reduction via elimination or Householder transformations to the upper Hessenberg form is undertaken, resulting in a simpler matrix form. The next, decisive step comprises the actual solution, which is obtained iteratively, since we are essentially dealing with a root computation. One especially mention-worthy method of determining all the eigenvalues and eigenvectors is the QR method described in detail by Wilkinson [64], which works even with several eigenvalues. If all the similarity transformations executed during the calculation are stored and then canceled again, we then obtain the eigenvectors of the problem as well.

The particular solution according to (9.17) is computed numerically simply by means of matrix multiplications and additions. The complex frequency response vector (9.19) is determined with the linear system of equations

$$(ik\Omega \boldsymbol{E} - \boldsymbol{A}) \cdot (\boldsymbol{g}_k^{(1)} - i\boldsymbol{g}_k^{(2)}) = \boldsymbol{b}_k^{(1)} - i\boldsymbol{b}_k^{(2)}, \qquad k = 1(1)n. \tag{9.27}$$

Gaussian elimination with a column pivot is an example of a numerical method that can be used in this context. However, solving (9.27) is only recommendable if there is a fixed excitation frequency Ω. If we were to look for the frequency response vector as a function of the excitation frequency in the case of a harmonic excitation, $l = 1$, it would be more advisable to resort to the elementary frequency responses with known solutions as shown in Müller and Schiehlen [37]. The elementary frequency responses again require solving the eigenvalue problem.

The Ljapunov matrix (9.26) equation can be solved either by converting it into a higher-order system of linear equations or iteratively with the Smith method [60]. The Smith method is numerically more advantageous because of the unchanged order. The convergence of the iteration is constant due to the required asymptotic stability of the system.

The existence of algebraic solutions for the time-invariant state equation (9.11) of course does not exclude the possibility of utilizing numerical integration methods for linear systems as well. This approach is actually taken very often in practice for the sake of convenience. Yet we should always be aware that the initial value problem can only describe dynamic behavior for a single initial state, while the eigenvalue problem contains the entire range of solutions for any initial conditions.

All the numerical methods mentioned are also available in Matlab, todays standard mathematical toolbox for engineering and scientific computations in industry and universities.

9.3 Comparison of Mechanical Models

In order to compare numerically the models of multibody systems, finite element systems, and continuous systems, only simple designs are suitable that are accessible to a closed solution. We will therefore take as an example the longitudinal vibrations of a bar, see Fig. 9.1. The parameters of a homogeneous bar fixed on one side include density ρ, cross-sectional area A, Young's modulus E, and length L.

In accordance with the method of multibody systems, the bar will be described in the following by f mass points, where f denotes the number of degrees of freedom. Then the equations of motion are written according to (5.60) e.g. for $f = 4$ mass points

$$\frac{\rho AL}{8}\begin{bmatrix} 2 & 0 & 0 & 0 \\ 0 & 2 & 0 & 0 \\ 0 & 0 & 2 & 0 \\ 0 & 0 & 0 & 1 \end{bmatrix} \cdot \begin{bmatrix} \ddot{u}_1 \\ \ddot{u}_2 \\ \ddot{u}_3 \\ \ddot{u}_4 \end{bmatrix} + \frac{4AE}{L}\begin{bmatrix} 2 & -1 & 0 & 0 \\ -1 & 2 & -1 & 0 \\ 0 & -1 & 2 & -1 \\ 0 & 0 & -1 & 1 \end{bmatrix} \cdot \begin{bmatrix} u_1 \\ u_2 \\ u_3 \\ u_4 \end{bmatrix} = \mathbf{0}. \qquad (9.28)$$

Half of the continuously distributed mass of the bar is added to each mass point, and the spring constants are determined using Hooke's law.

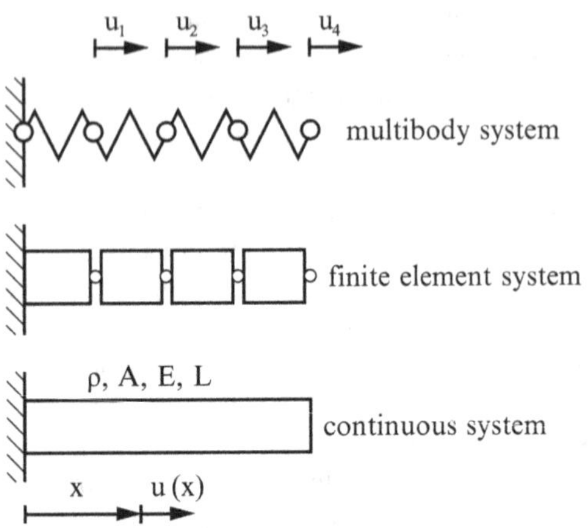

Fig. 9.1 Models for the longitudinal vibrations of a bar

The method of finite elements yields, according to (6.23) and (6.28), for four elements

$$\frac{\rho AL}{24}\begin{bmatrix}4&1&0&0\\1&4&1&0\\0&1&4&1\\0&0&1&2\end{bmatrix}\cdot\begin{bmatrix}\ddot{u}_1\\\ddot{u}_2\\\ddot{u}_3\\\ddot{u}_4\end{bmatrix}+\frac{4AE}{L}\begin{bmatrix}2&-1&0&0\\-1&2&-1&0\\0&-1&2&-1\\0&0&-1&1\end{bmatrix}\cdot\begin{bmatrix}u_1\\u_2\\u_3\\u_4\end{bmatrix}=\mathbf{0}. \quad (9.29)$$

The continuous systems lead, with (7.18) under the given boundary conditions, to the eigenvalue problem

$$\cos\beta L=0 \quad (9.30)$$

with the eigenfrequencies

$$\omega_f=\frac{(2f-1)\pi}{2}\sqrt{\frac{E}{\rho L^2}}, \qquad f=1(1)\infty. \quad (9.31)$$

As an initial, very rough test, the eigenfrequencies for $f=1$ will be compared. The exact value is obtained from (9.31)

$$\omega_1=1.570\sqrt{\frac{E}{\rho L^2}}. \quad (9.32)$$

The method of multibody systems yields for one mass point

$$\omega_1=1.414\sqrt{\frac{E}{\rho L^2}} \quad (9.33)$$

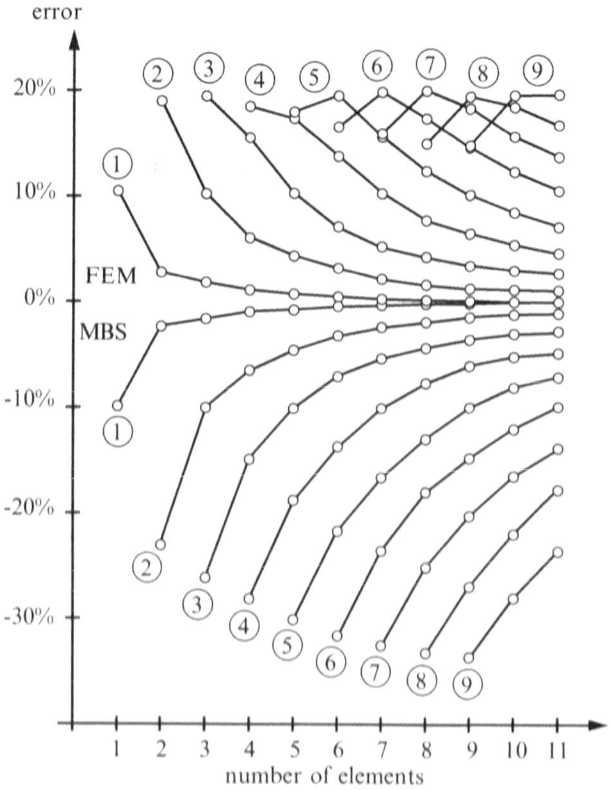

Fig. 9.2 Frequency error as a function of the number of elements

and the finite element method provides for one finite element

$$\omega_1 = 1.732\sqrt{\frac{E}{\rho L^2}}. \tag{9.34}$$

We first notice that all methods exhibit the correct dependence of the bar parameters. Only the numerical factor differs. The frequency error of the method of multibody systems amounts to $-10\,\%$, the method of finite elements has a frequency error of $+10\,\%$. For a rough approximation with $f = 1$, this is already an astonishingly good result!

We can now increase the number of degrees of freedom and plot the frequency errors referring to the exact frequencies (9.31), see Fig. 9.2. The method of multibody systems here invariably provides too small frequencies, while the method of finite elements results in too large ones. This can be easily explained. The concentration of the distributed mass in the mass points magnifies the effects of inertia, which must reduce the frequency. On the other hand, the linear distribution of mass in the method of finite elements reduces the effects of inertia, leading to higher frequencies. Also, Fig. 9.2 shows that the error decreases with an increasing number of degrees of freedom. Already with three degrees of freedom, the error of the first eigenfrequency is only $\pm 1\,\%$. This fact among others is the reason for the great success enjoyed by the methods of multibody systems and of finite elements. In general, it can be said that the finite element method leads to more accurate results in the case of linear systems, as is shown in Fig. 9.2. With a sufficient number of degrees of freedom, the method of multibody systems can attain the accuracy of the finite element method in this case too. It is also excellently applicable to large, nonlinear motions.

The basic application areas of the finite element method are in structural dynamics, while the method of multibody systems is primarily used in machine dynamics. Continuous systems are widely utilized in theoretical dynamics.

Appendix A
Mathematical Tools

This appendix compiles several frequently used relations and definitions.

A.1 Representation of Functions

In order to represent functions, a simplified notation is used. This will be illustrated in the following using the example of the location vector of a mass point. The free motion of a mass point takes place in three-dimensional Euclidean space R^3. Its position is then described with a 3×1 location vector $\boldsymbol{r}$,

$$\boldsymbol{r} \in R^3. \tag{A.1}$$

If the motion is limited by a holonomic constraint to a surface, then a position vector $\boldsymbol{y} \in R^2$ of the generalized coordinates suffices in order to determine its position uniquely. Then we can understand the location vector as a function of the vector $\boldsymbol{y}$,

$$\boldsymbol{r} = \boldsymbol{f}(\boldsymbol{y}), \qquad \boldsymbol{f} : R^2 \to R^3. \tag{A.2}$$

This means that each element $\boldsymbol{y}$ of the configuration space R^f is assigned exactly one element $\boldsymbol{r}$ of the three-dimensional space R^3 by the mapping $\boldsymbol{f}$.

The basic equations of kinetics yield together with an initial condition

$$\boldsymbol{y}_0 \in R^f, \qquad \dot{\boldsymbol{y}}_0 \in R^f \tag{A.3}$$

the position vector $\boldsymbol{y}$ as a function of time t,

$$\boldsymbol{y} = \boldsymbol{g}(t), \qquad \boldsymbol{g} : R \to R^f. \tag{A.4}$$

Thus, from (A.2) follows the dependence of the location vector on time

$$\boldsymbol{r} = \boldsymbol{f}(\boldsymbol{y}) = \boldsymbol{f}(\boldsymbol{g}(t)) = \boldsymbol{h}(t). \tag{A.5}$$

W. Schiehlen and P. Eberhard, *Applied Dynamics*, DOI 10.1007/978-3-319-07335-4,

As we can see, several mapping steps are required in order to describe exactly even the simplest of mechanical models. The resultantly complex notation is thus replaced with a simplified description. To this end, (A.1)–(A.5) are merged into a single relation

$$\boldsymbol{r}(t) = \boldsymbol{r}(\boldsymbol{y}(t)) = \boldsymbol{r}(\boldsymbol{y}). \tag{A.6}$$

For the sake of a concise notation, we thus consciously neglect the fact that $\boldsymbol{f}$ and $\boldsymbol{h}$ are two different mappings. The resulting reduction of the text justifies the inaccuracies that result from the abbreviated notation.

A.2 Matrix Algebra

The rectangular $m \times n$ matrix $\boldsymbol{A}$ is formed by the pattern of its elements $A_{\alpha\beta}$,

$$\boldsymbol{A} = \left[A_{\alpha\beta}\right], \qquad \alpha = 1(1)m, \qquad \beta = 1(1)n. \tag{A.7}$$

The matrix addition of two $m \times n$ matrices $\boldsymbol{A}$ and $\boldsymbol{B}$ is undertaken element by element

$$\boldsymbol{A} + \boldsymbol{B} = \left[A_{\alpha\beta} + B_{\alpha\beta}\right] = \left[C_{\alpha\beta}\right] = \boldsymbol{C}. \tag{A.8}$$

The matrix multiplication of an $m \times p$ matrix $\boldsymbol{A}$ and a $p \times n$ matrix $\boldsymbol{B}$ is

$$\boldsymbol{A} \cdot \boldsymbol{B} = \left[\sum_{\gamma=1}^{p} A_{\alpha\gamma} B_{\gamma\beta}\right] = [C_{\alpha\beta}] = \boldsymbol{C}. \tag{A.9}$$

In compliance with the Einstein summation convention, the summation sign can be omitted. We should then always sum over the indices appearing doubled in one term.

The transposed $n \times m$ matrix

$$\boldsymbol{A}^T = \left[A_{\beta\alpha}\right] \tag{A.10}$$

is obtained by exchanging columns and rows. The square $n \times n$ matrix $\boldsymbol{A}$ can be regular

$$\det \boldsymbol{A} \neq 0 \tag{A.11}$$

or singular

$$\det \boldsymbol{A} = 0. \tag{A.12}$$

For every regular matrix there exists an $n \times n$ inverse matrix $\boldsymbol{A}^{-1}$, which satisfies the condition

$$\boldsymbol{A} \cdot \boldsymbol{A}^{-1} = \boldsymbol{E}. \tag{A.13}$$

The regular $n \times n$ matrix $\boldsymbol{A}$ is referred to as orthogonal if it satisfies the condition

$$\boldsymbol{A}^T = \boldsymbol{A}^{-1}, \qquad \det\boldsymbol{A} = \pm 1. \tag{A.14}$$

Also, for

$$\det\boldsymbol{A} = 1, \tag{A.15}$$

the $n \times n$ matrix $\boldsymbol{A}$ is proper orthogonal.

Every square $n \times n$ matrix can be additively dismantled uniquely into a symmetric and a skew-symmetric matrix

$$\boldsymbol{A} = \frac{1}{2}(\boldsymbol{A} + \boldsymbol{A}^T) + \frac{1}{2}(\boldsymbol{A} - \boldsymbol{A}^T) = \boldsymbol{B} + \boldsymbol{C}. \tag{A.16}$$

The symmetric $n \times n$ matrix $\boldsymbol{B}$ satisfies the condition

$$\boldsymbol{B} = \boldsymbol{B}^T \tag{A.17}$$

and the skew-symmetric $n \times n$ matrix $\boldsymbol{C}$ is characterized by the relations

$$\boldsymbol{C} = -\boldsymbol{C}^T, \qquad C_{\alpha\alpha} = 0, \qquad \alpha = 1(1)n. \tag{A.18}$$

A symmetric matrix has $\frac{n}{2}(n+1)$ essential elements, while a skew-symmetric matrix only has $\frac{n}{2}(n-1)$ essential elements. It is sometimes helpful to summarize the essential elements in one vector of corresponding dimensions.

The symmetric $n \times n$ matrix $\boldsymbol{A}$ is called positive definite if the quadratic form

$$\boldsymbol{x} \cdot \boldsymbol{A} \cdot \boldsymbol{x} > 0 \qquad \forall \boldsymbol{x} \neq \boldsymbol{0} \tag{A.19}$$

and it is positive semidefinite for

$$\boldsymbol{x} \cdot \boldsymbol{A} \cdot \boldsymbol{x} \geq 0 \qquad \forall \boldsymbol{x}. \tag{A.20}$$

In order to test definiteness via computation, the principal minors H_α can also be exploited. The matrix $\boldsymbol{A}$ is positive definite for $H_\alpha > 0$ and positive semidefinite for $H_\alpha \geq 0, \alpha = 1(1)n$.

The $n \times n$ diagonal matrix

$$\boldsymbol{A} = \boldsymbol{diag}\{A_{\alpha\alpha}\}, \qquad A_{\alpha\beta} = 0 \qquad \text{for } \alpha \neq \beta \tag{A.21}$$

is only occupied on the principal diagonals. The essential elements of an $n \times n$ diagonal matrix can easily be written in an $n \times 1$ vector.

The $n \times n$ unit matrix

$$\boldsymbol{E} = \boldsymbol{diag}\{E_{\alpha\alpha}\}, \qquad E_{\alpha\beta} = \begin{cases} 1 & \text{for } \alpha = \beta \\ 0 & \text{otherwise} \end{cases} \tag{A.22}$$

is an proper orthogonal, positive definite diagonal matrix.

Every square 3×3 matrix $\boldsymbol{A}$ can be decomposed in a polar fashion into an proper orthogonal matrix $\boldsymbol{B}$ and positive definite matrices $\boldsymbol{C}, \boldsymbol{D}$,

$$\boldsymbol{A} = \boldsymbol{B} \cdot \boldsymbol{C} = \boldsymbol{D} \cdot \boldsymbol{B} \tag{A.23}$$

with

$$\boldsymbol{C} = (\boldsymbol{A}^T \cdot \boldsymbol{A})^{\frac{1}{2}}, \qquad \boldsymbol{D} = (\boldsymbol{A} \cdot \boldsymbol{A}^T)^{\frac{1}{2}} \tag{A.24}$$

and

$$\boldsymbol{B} = \boldsymbol{A} \cdot \boldsymbol{C}^{-1} = \boldsymbol{D}^{-1} \cdot \boldsymbol{A}. \tag{A.25}$$

It should be noted that for every positive definite 3×3 matrix $\boldsymbol{C}$ an also positive definite 3×3 matrix $\boldsymbol{C}^{1/2}$ can be found. They are calculated with a principal axis transformation, see Zurmühl and Falk [69] or Golub and van Loan [23].

The skew-symmetric 3×3 matrix

$$\tilde{\boldsymbol{a}} = \sum_{\beta=1}^{3} \varepsilon_{\alpha\beta\gamma} a_{\beta}, \qquad \alpha, \gamma = 1(1)3 \tag{A.26}$$

is uniquely determined by the 3×1 vector

$$\boldsymbol{a} = \left[a_{\alpha} \right], \qquad \alpha = 1(1)3. \tag{A.27}$$

The permutation symbol $\varepsilon_{\alpha\beta\gamma}$ has the properties

$$\begin{aligned} &\varepsilon_{\alpha\beta\gamma} = \varepsilon_{\beta\gamma\alpha} = \varepsilon_{\gamma\alpha\beta} = 1, \\ &\varepsilon_{\beta\alpha\gamma} = \varepsilon_{\gamma\beta\alpha} = \varepsilon_{\alpha\gamma\beta} = -1, \\ &\varepsilon_{\alpha\beta\gamma} = 0 \qquad \text{for } \alpha = \beta, \beta = \gamma \text{ or } \gamma = \alpha. \end{aligned} \tag{A.28}$$

The skew-symmetric 3×3 matrix gives in particular the vector product of two vectors

$$\boldsymbol{a} \times \boldsymbol{b} = \tilde{\boldsymbol{a}} \cdot \boldsymbol{b} = \left[\sum_{\beta,\gamma=1}^{3} \varepsilon_{\alpha\beta\gamma} a_{\beta} b_{\gamma} \right] = [c_{\alpha}] = \boldsymbol{c}. \tag{A.29}$$

The following relations remain valid

$$\tilde{\boldsymbol{a}} \cdot \tilde{\boldsymbol{b}} = \boldsymbol{ba} - (\boldsymbol{a} \cdot \boldsymbol{b})\boldsymbol{E}, \tag{A.30}$$

$$\widetilde{(\tilde{\boldsymbol{a}} \cdot \boldsymbol{b})} = \boldsymbol{ba} - \boldsymbol{ab}, \tag{A.31}$$

which express the multiple vector products in matrix notation.

In contrast, $\boldsymbol{ba}$ is the dyadic product of two vectors, which leads to the symmetric matrix

$$\boldsymbol{C} = [C_{\alpha\beta}] = [b_\alpha a_\beta] = [a_\beta b_\alpha] = [C_{\beta\alpha}] = \boldsymbol{C}^T. \tag{A.32}$$

Furthermore, the scalar product of two 3×1 vectors is

$$\boldsymbol{a} \cdot \boldsymbol{b} = \sum_{i=1}^{3} a_i b_i = c. \tag{A.33}$$

The symbol '$\cdot$' thus conveys both the matrix product (A.9) and the scalar product.

Column and row vectors are basically not distinguished in this book. When using subvectors, usually row notation is utilized, e.g.,

$$\boldsymbol{a} = [\boldsymbol{b} \quad \boldsymbol{c}] = [b_1 \quad b_2 \quad b_3 \quad \ldots \quad c_1 \quad c_2 \quad c_3 \quad \ldots]. \tag{A.34}$$

The relations applying to the 3×1 vectors and 3×3 matrices can be transferred directly to tensors of the 1st and 2nd order.

A.3 Matrix Analysis

The differentiation and integration of a rectangular $m \times n$ matrix $\boldsymbol{A}$ according to a scalar quantity t is carried out element by element

$$\frac{d\boldsymbol{A}}{dt} = \left[\frac{dA_{\alpha\beta}}{dt}\right], \qquad \alpha = 1(1)m, \qquad \beta = 1(1)n. \tag{A.35}$$

The partial differentiation of an $m \times 1$ vector $\boldsymbol{a}(\boldsymbol{b})$ according to an $n \times 1$ vector $\boldsymbol{b}$ leads to a $m \times n$ functional or Jacobian matrix

$$\frac{\partial \boldsymbol{a}}{\partial \boldsymbol{b}} = \left[\frac{\partial a_\alpha}{\partial b_\beta}\right] = [C_{\alpha\beta}] = \boldsymbol{C}, \qquad \alpha = 1(1)m, \qquad \beta = 1(1)n. \tag{A.36}$$

In matrix notation, the $m \times 1$ vector $\boldsymbol{a}$ of the dependent variables is differentiated with respect to the $n \times 1$ vector $\boldsymbol{b}$ of the independent variables, as shown in (A.36). An example of the above definition can be found in (2.9).

According to the chain rule, the differentiations can also be applied multiple times, which leads to a corresponding multiplication of the Jacobian matrices, see (2.10) and (2.11).

According to (A.36), with a 3×1 location vector $\boldsymbol{x}$ the notations of vector analysis can also be written

$$\text{grad}\boldsymbol{a} = \frac{\partial \boldsymbol{a}}{\partial \boldsymbol{x}}, \qquad \text{div}\boldsymbol{a} = \text{Sp}\,\frac{\partial \boldsymbol{a}}{\partial \boldsymbol{x}}. \tag{A.37}$$

This summarizes the definitions of matrix analysis which are essential for applied dynamics.

A.4 List of Important Symbols

Many symbols have different meanings in individual chapters, which is due to the interdisciplinary nature of applied dynamics. Therefore, the number of the formula in which the symbol first appears is added in the brackets.

a	Constant (6.44)
$\boldsymbol{a}$	3×1 acceleration vector (2.12)
$\boldsymbol{a}$	$n \times 1$ vector function (8.4)
b	Constant (6.44)
b	Distance (2.22)
$\boldsymbol{b}$	$n \times 1$ excitation function (9.11)
$\boldsymbol{b}$	6×1 vector of acceleration quantities (5.167)
c	Spring constant (4.22)
d	Damping constant (5.92)
$\boldsymbol{d}$	3×1 vector of the rotation axis (2.36)
e	Number of degrees of freedom (1.1)
$\boldsymbol{e}$	Unit vector (2.1)
$\boldsymbol{e}$	6×1 strain vector (2.141)
$\boldsymbol{e}$	Position vector elastic beam (6.59)
f	Number of position degrees of freedom (2.174)
$\boldsymbol{f}$	3×1 force vector (3.1)
$\boldsymbol{f}$	3×1 vector of inertia force density (3.61)
$\boldsymbol{f}$	$f \times 1$ vector function (5.39)
g	Number of velocity degrees of freedom (2.218)
g	Gravitational acceleration (3.17)
$\boldsymbol{g}$	$q \times 1$ vector of generalized reaction forces (3.8)
$\boldsymbol{g}$	$n \times 1$ frequency response vector (9.20)
h	Number of force variables (3.12)

$\boldsymbol{h}$	$f \times 1$ excitation vector (5.59)
k	Inertia parameter (3.54)
$\boldsymbol{k}$	$f \times 1$ vector of generalized Coriolis, centrifugal, or gyroscopic forces (5.28)
l	Order of a series expansion (9.14)
$\boldsymbol{l}$	3×1 torque vector (3.23)
m	Mass (3.1)
$\boldsymbol{m}$	$n \times 1$ mean vector (9.21)
n	Distributed load on bar (7.7)
$\boldsymbol{n}$	3×1 normal vector (3.62)
p	Number of points, elements, bodies (2.24)
q	Quaternions (2.44)
q	Number of holonomic constraints (2.174)
q	Distributed load on beam (7.10)
$\boldsymbol{q}$	$f \times 1$ vector of generalized applied forces (5.28)
$\bar{\boldsymbol{q}}$	Global $e \times 1$ force vector (5.23)
R	Spherical coordinate (2.14)
r	Number of nonholonomic constraints (2.218)
r	Number of input variables (8.3)
$\boldsymbol{r}$	3×1 location vector (2.1)
ds	3×1 vector of infinitesimal rotation (2.85)
t	Time (2.1)
$\boldsymbol{t}$	3×1 stress vector (3.33)
u	Longitudinal displacement of the beam (6.12)
u	Control force (8.18)
$\boldsymbol{u}$	3×1 distance vector (3.34)
$\boldsymbol{u}$	$r \times 1$ input vector (8.3)
v	Deflection of the beam in the 2-direction (6.12)
$\boldsymbol{v}$	3×1 velocity vector (2.5)
w	Deflection of the beam in the 3-direction (6.12)
$\boldsymbol{w}$	3×1 displacement vector (2.134)
$\boldsymbol{w}$	$h \times 1$ vector of force variables (3.12)
$\boldsymbol{w}$	6×1 twist (5.166)
x	Coordinate of the beam axis (6.20)
$\boldsymbol{x}$	$e \times 1$ position vector of the free system (2.3)
$\boldsymbol{x}$	$n \times 1$ state vector (8.1)
$\boldsymbol{y}$	$f \times 1$ position vector of the holonomic system (2.177)
$\boldsymbol{z}$	$g \times 1$ velocity vector of the nonholonomic system (2.221)
A	Area, surface, cross-sectional area (3.33), (6.23)
$\boldsymbol{A}$	$3 \times f$ matrix of relative shape functions (2.148)
$\boldsymbol{A}$	$n \times n$ system matrix (8.5)
$\boldsymbol{B}$	$6 \times f$ matrix of stress functions (2.149)

$\boldsymbol{B}$	$n \times r$ input matrix (8.5)
$\boldsymbol{C}$	$3 \times f$ matrix of absolute shape functions (2.151)
$\boldsymbol{D}$	3×3 strain velocity tensor (2.156)
$\boldsymbol{D}$	$f \times f$ damping matrix (5.60)
$\mathscr{D}$	3×3 differential operator matrix of rotation (2.146)
E	Young's modulus (3.69)
$\boldsymbol{E}$	Unit matrix (2.9)
$\boldsymbol{F}$	3×3 deformation gradient (2.28)
$\boldsymbol{F}$	$3 \times q$ distribution matrix of the reaction forces (3.9)
G	Shear modulus (6.32)
$\boldsymbol{G}$	3×3 Green strain tensor (2.126)
$\boldsymbol{G}$	$f \times f$ matrix of gyroscopic forces (5.60)
$\boldsymbol{G}$	$e \times q$ functional matrix (4.13)
$\boldsymbol{G}$	$6 \times 6p$ summation matrix (5.141)
$\boldsymbol{H}$	6×6 Hooke matrix (3.68)
$\overline{\boldsymbol{H}}$	$e \times e$ Global Jacobian matrix (5.21)
$\boldsymbol{H}_{T,R}$	$3 \times e$ Jacobian matrix (2.6)
$\boldsymbol{I}$	$e \times f$ functional matrix (2.189)
$\boldsymbol{I}$	3×3 inertia tensor (3.29)
J	Area moment of inertia (5.128)
$\overline{\boldsymbol{J}}$	$e \times f$ Global Jacobian matrix (5.18)
$\boldsymbol{J}_{T,R}$	$3 \times f$ Jacobian matrix (2.184)
$\boldsymbol{K}$	$f \times g$ functional matrix (2.233)
$\boldsymbol{K}$	$f \times f$ stiffness matrix (5.60)
L	Length (6.23)
L	Lagrange function (4.50)
$\boldsymbol{L}$	3×3 velocity gradient (2.155)
$\boldsymbol{L}$	$3 \times q$ distribution matrix of the reaction torques (3.41)
$\overline{\boldsymbol{L}}$	Global $e \times g$ Jacobian matrix (5.21)
$\boldsymbol{L}_{T,R}$	$3 \times g$ Jacobian matrix (2.229)
M	Bending torque (6.37)
$\boldsymbol{M}$	Inertia matrix (5.29)
N	Normal force on the bar (7.9)
$\boldsymbol{N}$	6×3 matrix to the normal vector (4.35)
$\boldsymbol{N}$	$f \times f$ matrix of circulatory forces (5.60)
$\boldsymbol{N}$	Reaction matrix (5.107)
$\boldsymbol{N}$	$n \times n$ correlation matrix (9.23)
$\boldsymbol{O}$	Zero matrix (2.178)
$\boldsymbol{P}$	$f \times f$ matrix of velocity forces (5.59)
$\boldsymbol{P}$	$n \times n$ covariance matrix (9.25)
Q	Shear force (6.37)

$\boldsymbol{Q}$	4×4 coefficient matrix (2.92)
$\boldsymbol{Q}$	$f \times f$ matrix of the position forces (5.59)
$\underline{\boldsymbol{Q}}$	$n \times n$ intensity matrix (9.21)
$\overline{\boldsymbol{Q}}$	Global $e \times q$ distribution matrix (5.18)
R	Rayleigh function (5.61)
$\boldsymbol{S}$	3×3 rotation tensor (2.33)
$\overline{\boldsymbol{S}}$	3×3 rotation tensor (2.123)
T	Kinetic energy (4.50)
$\boldsymbol{T}$	3×3 stress tensor (3.62)
$\boldsymbol{T}$	$n \times n$ transformation matrix (8.7)
U	Potential energy (4.37)
U	Eigenfunction of a bar (7.18)
$\boldsymbol{U}$	3×3 right-stretch tensor (2.123)
$\boldsymbol{U}$	$f \times f$ transformation matrix (8.13)
V	Volume (3.18)
$\mathscr{V}$	6×3 strain operator matrix (2.142)
W	Work (4.1)
$\boldsymbol{W}$	3×3 rotational velocity tensor (2.156)
$\mathscr{W}$	3×6 differential operator matrix (7.5)
$\boldsymbol{X}$	$n \times n$ modal matrix (8.28)
$\boldsymbol{Y}$	$f \times f$ modal matrix (8.36)
α	Cardano angle (2.55)
$\boldsymbol{\alpha}$	3×1 rotational acceleration vector (2.118)
β	Cardano angle (2.56)
β	Frequency (7.20)
γ	Cardano angle (2.57)
γ	Shear stress (2.140)
γ	Slope of beam (6.56)
γ	Frequency (7.30)
δ	Virtual size (2.186)
δ	Rotation angle (6.75)
δ	Dirac function (7.28)
ε	Integration error (2.106)
ε	Strain (2.140)
ε	Permutation symbol (A.28)
ζ	6×1 vector of generalized functions (7.5)
$\boldsymbol{\eta}$	Small $f \times 1$ position vector (2.278)
ϑ	Spherical coordinate, Euler angle (2.14), (2.58)
λ	Eigenvalue (2.50)
ν	Poisson ratio (3.69)
ν	Eigenfrequency (5.49)

ρ	Density (3.18)
$\boldsymbol{\rho}$	3×1 location vector (2.25)
σ	Angle between axes (2.34)
σ	Normal stress (3.65)
$\boldsymbol{\sigma}$	6×1 stress vector (3.66)
τ	Shear stress (3.65)
φ	Rotation angle (2.36)
φ	Euler angle (2.58)
$\boldsymbol{\phi}$	$q \times 1$ vector of holonomic constraints (2.175)
ψ	Spherical coordinate, Euler angle (2.14), (2.58)
$\boldsymbol{\psi}$	$r \times 1$ vector of nonholonomic constraints (2.219)
$\boldsymbol{\omega}, \tilde{\boldsymbol{\omega}}$	3×1 rotation velocity vector/tensor (2.85)
$\boldsymbol{\phi}$	$n \times n$ fundamental matrix (9.12)
Ω	Rotation velocity (2.270)
Ω	Excitation frequency (5.134)
$\boldsymbol{\Omega}$	$f \times f$ frequency matrix (7.37)

The symbols are accompanied by the following often recurring indices.

a	External force (3.5)
c	Coriolis force (5.1)
e	Applied force (3.5)
e	Elastic displacement (6.72)
i	Internal force (3.5)
i	Element number, $i = 1(1)p$, (2.24)
j	Element number, reference system number, $j = 1(1)n$ (2.260)
n	Quaternion number, $n = 0(1)3$, (2.44)
p	Particular solution (5.50)
r	Reaction force (3.5)
s	Rigid body displacement (6.72)
w	Linear displacement (2.136)
H	Principal axes frame (3.51)
I	Inertial frame (2.248)
K	Rigid body (2.249)
O	Reference configuration (2.25)
P	Mass point (2.79)
P	Polar area moment of inertia (5.128)
R	Reference framen (2.247)
R	Rotation (2.100)
S	Target motion (2.278)
T	Transposed matrix (2.79)
T	Translation (2.7)

α	Direction of the base vector, $\alpha = 1(1)3$, (2.1)
β	Direction of the base vector, $\beta = 1(1)3$, (2.34)
γ	Number of the generalized coordinate, $\gamma = 1(1)3$, (2.3)

Bibliography

1. Arnold VI (1989) Mathematical methods of classical mechanics. Springer, New York
2. Arnold R, Maunder L (1961) Gyrodynamics and its engineering applications. Academic Press, New York
3. Arnold M, Schiehlen W (eds) (2008) Simulation techniques for applied dynamics. Springer, Wien
4. Bae DS, Haug EJ (1987) A recursive formulation for constrained mechanical system dynamics: part I, open loop systems. Mech Struct Mach 15:359–382
5. Bae DS, Haug EJ (1987) A recursive formulation for constrained mechanical system dynamics: part II, closed loop systems. Mech Struct Mach 15:481–506
6. Bathe KJ (1995) Finite element procedures. Prentice Hall, Upper Saddle River
7. Bauchau OA (2010) Flexible multibody dynamics. Springer, Dordrecht
8. Bauchau O, Bottasso C, Nikishkov Y (2001) Modelling rotorcraft dynamics with finite element multibody procedures. Math Comput Model 33:1113–1137
9. Becker E, Bürger W (1975) Kontinuumsmechanik. Teubner, Stuttgart
10. Biezeno CB, Grammel R (1971) Technische Dynamik, 2vols. Springer, Berlin
11. Brandl H, Johanni R, Otter M (1988) A very efficient algorithm for the simulation of robots and similar multibody systems without inversion of the mass matrix. In: Kopacek P, Troch I, Desoyer K (eds) Theory of robots. Pergamon, Oxford, pp 95–100
12. Bronstein IN, Semendjajew KA, Musiol G, Mühlig H (2013) Taschenbuch der Mathematik. Harri Deutsch, Frankfurt a.M.
13. Budo A (1990) Theoretische Mechanik. Deutscher Verlag der Wissenschaften, Berlin
14. Butcher JC (2008) Numerical methods for ordinary differential equations. Wiley-Blackwell, Hoboken
15. Demeter GF (1995) Mechanical and structural vibrations. Wiley, Hoboken
16. Dresig H, Holzweissig F (2012) Maschinendynamik. Springer, Berlin
17. Eberhard P (2000) Kontaktuntersuchungen durch hybride Mehrkörpersystem/Finite Elemente Simulationen. Shaker Verlag, Aachen
18. Eich-Soellner E, Führer C (2013) Numerical methods in multibody dynamics. Vieweg Teubner, Wiesbaden
19. Escalona J, Hussien H, Shabana A (1998) Application of the absolute nodal coordinate formulation to multibody system dynamics. J Sound Vib 214(5):833–851
20. Euler L (1775) Novi commentarii academiae scientiarum petropolitanae. Academia Scientiarum, St. Petersburg
21. Fehr J, Eberhard P (2011) Simulation process of flexible multibody systems with advanced model order reduction techniques. Multibody Syst Dyn 25(3):313–334
22. Gerardin M, Cardona A (2001) Flexible multibody dynamics – a finite element approach. Wiley, Chichester

W. Schiehlen and P. Eberhard, *Applied Dynamics*, DOI 10.1007/978-3-319-07335-4,

23. Golub G, van Loan C (2013) Matrix computations. John Hopkins University Press, Baltimore
24. Grigorieff RD (1972, 1977) Numerik gewöhnlicher Differentialgleichungen, 2vols. Teubner, Stuttgart
25. Hamel G (2013) Theoretische Mechanik. Springer, Berlin
26. Hiller M (1983) Mechanische Systeme. Springer, Berlin
27. Hollerbach JM (1980) A recursive Lagrangian formulation of manipulator dynamics and comparative study of dynamics formulation complexity. IEEE Trans Syst Man Cybern 11:730–736
28. Irretier H (2000) Grundlagen der Schwingungstechnik 1/2. Vieweg, Braunschweig
29. Kreuzer E, Schiehlen W (1985) Computerized generation of symbolic equations of motion for spacecraft. J Guid Control Dyn 8(2):284–287
30. Kurz T, Eberhard P, Henninger C, Schiehlen W (2010) From Neweul to Neweul-M2: symbolical equations of motion for multibody system analysis and synthesis. Multibody Syst Dyn 24(1):25–41
31. Lai W, Rubin D, Krempl E (2009) Introduction to continuum mechanics. Butterworth Heinemann, Oxford
32. Lehner M, Eberhard P (2006) On the use of moment matching to build reduced order models in flexible multibody dynamics. Multibody Syst Dyn 16(2):191–211
33. Link M (2014) Finite Elemente in der Statik und Dynamik. Springer, Wiesbaden
34. Lurie AI (2002) Analytical mechanics. Springer, Berlin
35. Magnus K (1971) Kreisel – Theorie und Anwendungen. Springer, Berlin
36. Magnus K, Müller-Slany HH (2005) Grundlagen der Technischen Mechanik. Teubner, Stuttgart
37. Müller PC, Schiehlen WO (2005) Linear vibrations. Springer, Dordrecht
38. Nayfeh A, Mook D (2005) Nonlinear oscillations. Wiley, Hoboken
39. Newton I (1687) Philosophia Naturalis Principia Mathematica. Royal Society, London
40. Papastavridis JG (2002) Analytical mechanics. Oxford University Press, Oxford
41. Päsler M (1968) Prinzipe der Mechanik. de Gruyter, Berlin
42. Pfeiffer F, Glocker C (1996) Multibody dynamics with unilateral contacts. Wiley, New York
43. Popp K, Schiehlen W (1996) Ground vehicle dynamics. Springer, Berlin
44. Przemieniecki JS (1985) Theory of matrix structural analysis. Dover, New York
45. Rill G, Schaeffer T (2010) Grundlagen und Methodik der Mehrkörpersimulation mit Anwendungsbeispielen. Vieweg+Teubner, Wiesbaden
46. Rubin M (2010) Cosserat theories: shells, rods and points. Springer, Dordrecht
47. Saha SK, Schiehlen W (2001) Recursive kinematics and dynamics for parallel structural closed-loop multibody systems. Mech Struct Mach 29:143–175
48. Schäfer H (2001) Das Cosserat-Kontinuum. ZAMM 47(5):485–495
49. Schiehlen W (1991) Computational aspects in multibody system dynamics. Comput Methods Appl Mech Eng 90:569–582
50. Schiehlen W (1997) Multibody system dynamics: roots and perspectives. Multibody Syst Dyn 1:149–188
51. Schwertassek R, Wallrapp O (1997) Dynamik flexibler Mehrkörpersysteme. Vieweg, Wiesbaden
52. Seifried R (2013) Dynamics of underactuated multibody systems: modeling, control and optimal design. Springer, Heidelberg
53. Sextro W, Popp K, Magnus K (2013) Schwingungen: Eine Einführung in physikalische Grundlagen und die theoretische Behandlung von Schwingungsproblemen. Springer, Wiesbaden
54. Shabana AA (1996) Finite element incremental approach and exact rigid body inertia. ASME J Mech Des 118(2):171–178
55. Shabana AA (1997) Flexible multibody dynamics: review of past and recent developments. Multibody Syst Dyn 1(2):189–222
56. Shabana AA (1997) Dynamics of multibody systems. Cambridge University Press, Cambridge
57. Shabana AA (2010) Computational dynamics. Wiley, Chichester

58. Shampine LF, Gordon MK (2010) Computer solution of ordinary differential equations. Freeman, San Francisco
59. Simeon B (2013) Computational flexible multibody dynamics: a differential-algebraic approach. Springer, Heidelberg
60. Smith RA (2013) Matrix equation XA+BX=C. SIAM J Appl Math 16:198–201
61. Spivak M (2006) Calculus. Cambridge University Press, Cambridge
62. Steinke P (2010) Finite Elemente Methode: Rechnergestützte Einführung. Springer, Berlin
63. Szabo I (1977) Geschichte der mechanischen Prinzipien. Birkhäuser, Basel
64. Wilkinson JH (1977) The algebraic eigenvalue problem. Oxford University Press, Oxford
65. Wittenburg J (2008) Dynamics of multibody systems. Springer, Berlin
66. Woernle C (2011) Mehrkörpersysteme. Springer, Berlin
67. Wriggers P (2008) Nonlinear finite element methods. Springer, Berlin
68. Zienkiewicz OC, Taylor RL (2005) The finite element method (in 3 vols). Butterworth Heinemann, Oxford
69. Zurmühl R, Falk S (1997) Matrizen und ihre Anwendungen. Springer, Berlin

Index

W. Schiehlen and P. Eberhard, *Applied Dynamics*, DOI 10.1007/978-3-319-07335-4,

Printed in the United States
By Bookmasters